Introducing Palaeontology

Companion titles:

Introducing Geology – A Guide to the World of Rocks
(Second Edition 2010)

Introducing Volcanology – A Guide to Hot Rocks
(to be published 2011)

For further details on these and Dunedin's
other Earth Science and Geology publications see
www.dunedinacademicpress.co.uk

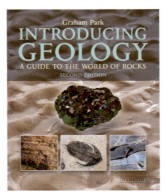

ISBN 978-1-906716-21-9

Introducing Palaeontology

A Guide to Ancient Life

Patrick N. Wyse Jackson

With illustrations by John Murray

DUNEDIN

For Mary

Published by
Dunedin Academic Press Ltd
Hudson House
8 Albany Street
Edinburgh EH1 3QB
Scotland

ISBN 978-1-906716-15-8
© 2010 Patrick N. Wyse Jackson

British Library Cataloguing in Publication Data
A catalogue record for this book is available from the British Library

Typeset by Makar Publishing Production, Edinburgh
Printed in Poland by Hussar Books

Contents

Acknowledgements

I have received advice and assistance from many quarters during the preparation of this book and in particular thank Charles Holland for his long-standing encouragement of my palaeontological endeavours. I am most grateful to a number of colleagues and friends who all kindly made images available for this book, and thank in particular John Murray (NUI Galway), whose artistic ability and palaeontological expertise have contributed greatly to this book. I thank Anne Morton and David McLeod for their help in preparing this book for the press, and Euan Clarkson for his helpful suggestions. Photographs or diagrams were supplied by José-Manuel Benito Álvarez (99B–C), Declan Burke (11, 44D, 70B), Neil Clark (79C), Geoffrey Clayton (32B, 43), Patricio Dominguez Alonso (70C), Gareth Dyke (37B, 94), Andrej Ernst (7B, 37A, C, 87), John Graham (102A), Sarah Heal (32A), Charles Holland (15B), Marcus Key (25A, 26, 52B), Jonathan Larwood/Natural England (9B), Alf Lenz (78B–D), John Murray (14, 23, 24, 28, 47A–B, 48D, 49, 51, 53, 55B–C, 57, 59, 62, 66, 68A–C, 69 left, 71, 77), Chris Nicholas (15A, 16B–C, 17, 18, 19, 76), John Nudds (16A, 39, 83A), Matthew Parkes (84), Colin Prosser/Natural England (9A), Jeremy Stone (79A–B), Adam Stuart Smith (88), Mark Wilson (3, 22, 35B–D, 36C, 42B, 48A, 50D, 58B, 60D, 61, 65, 67F, 102B), Patrick Wyse Jackson (1, 2, 5, 6, 7A, 8, 12, 13, 20, 21, 25B, 27, 31, 33, 34, 36A–B, 38, 39, 42A, C–E, 44A–C, 45, 48B–C, 50A–C, 52A. C–E, 54, 55, 56, 60A–C, 67A–E, 68D, 69 right, 70A, 72, 73, 74, 75A, C–D, 78A, 81, 82, 83B, 85, 89, 90, 91, 92, 93, 96, 97, 98, 103). I warmly thank Susanna Wyse Jackson who skilfully redrafted a number of figures.

Note on illustrations

Most of the specimens illustrated are in the collections of the Geological Museum, Trinity College, Dublin. Some specimens were photographed during visits to other museums which are indicated by the following acronyms: AMHM: American Museum of Natural History, New York; CMNHS: Cincinnati Museum of Natural History and Science; FOSP: Falls of the Ohio State Park Interpretive Center; JMMG: James Mitchell Museum, University College, Galway, LNHM: Lausanne Natural History Museum; TMP: Transvaal Museum, Pretoria. The dates given on the Geological Timescale (Figure 14) are those published in 2009.

Suggested additional reading and further information

The following books provide more detail on geology and many of the groups and topics covered in this book:

Benton, M. J. 2005. *Vertebrate Palaeontology*. Blackwell Publishing, Oxford.

Benton, M.J. and Harper, D.A.T. 2008. *Basic Palaeontology: Introduction to Palaeobiology and the Fossil Record*. Wiley Blackwell, London.

Briggs, D.E.G. and Crowther, P.R. (eds). 1990. *Palaeobiology: a synthesis*. Blackwell Publishing, Oxford.

Briggs, D.E.G. and Crowther, P.R. (eds). 2001. *Palaeobiology II*. Blackwell Publishing, Oxford.

Clarkson, E.N.K. 1998. *Invertebrate Palaeontology and Evolution*. Blackwell Publishing, Oxford.

Nudds, J.R. and Selden, P.A. 2007. *Fossil Ecosystems of North America: A guide to the sites and their extraordinary biotas*. Manson Publishing, London.

Park, G. 2010. *Introducing Geology – A Guide to the World of Rocks*. Dunedin Academic Press, Edinburgh.

Selden, P.A. and Nudds, J.R. 2005. *Evolution of Fossil Ecosystems*. Manson Publishing, London.

Stringer, C. and Andrews, P. 2005. *The Complete World of Human Evolution*. Thames and Hudson, London.

In addition there are many good guides available from commercial publishers, which can be used profitably to identify your fossil finds. Various societies include the Palaeontological Association (www.palass. org) and the Paleontological Society (www. paleosoc.org). Many organisations also exist for those interested in individual fossil groups, amongst them the International Bryozoology Association (bryozoans), the Pander Society (conodonts), and Commission Internationale de Microflore du Paléozoique (palynology and spores). There are many specialist journals that publish research on fossils, including *Acta Palaeontologia Polonica*, *Alchingera*, *Journal of Paleontology*, *Lethaia*, *Palaeontology*, *Palaios*, and *Paleobiology*. The former two concentrate largely on systematics and taxonomy, while the latter two focus on functional morphology, ecology, and interactions between fossils and their environment. *Ichnos* publishes research on trace fossils. Many geological surveys, as well as local scientific societies and groups, publish valuable guides to local geology and palaeontology. Of these, the Fossil Focus leaflets published by the British Geological Survey are most attractive. Your local geological museum can also provide assistance to collectors and students alike.

Part 1

The science of palaeontology

1.1 Preface: the fascination of fossils

Fossils are the remains of plants and animals that lived in the past, and which are now frequently, but not necessarily always, preserved in stone. They may be found either as body fossils (Figure 1) which represent all or parts of the organism, or as trace fossils (Figure 2) which indicate the activities of the past organisms. Many years ago the term 'fossil' referred to anything that had been dug up, but now it is restricted to the remains of ancient organisms.

Roman coins, old leather boots, and deposits of lead and zinc are not fossils, but trilobites and dinosaur eggs are. The study of fossils is called palaeontology and scientists who study fossils are called palaeontologists. The subject is generally studied as part of geology, the science that investigates the structure and history of the Earth and now adjacent planets.

This book has been written with two groups of readers in mind. Many members of the

Figure 1 The ammonites *Promicroceras planicosta* and *Asteroceras obtusum* from the Lower Jurassic of Marston Magna, Somerset, England. These are preserved in their original mother-of-pearl shell, and the rock is known as Marston Marble.

Figure 2 *Gigandipus*, a dinosaur footprint in the Lower Jurassic Moenave Formation at the St George Dinosaur Discovery Site at Johnson Farm, southwestern Utah, USA.

general public will be aware of fossils; those of Missouri in the USA are proud that crinoids are their State Fossil, but I suspect few residents of that State have a real grasp of the significance and nature of fossils. It is hoped that this book will develop and enhance the fascination of fossils amongst the general public. The book is also intended for first and second-year university students taking courses in geology, biology and palaeontology.

This book provides a broad introduction to palaeontology by means of two distinct sections. The first part discusses aspects and uses of the subject through a discussion of

how to collect fossils responsibly, how to care for collections, and how to name and classify them. Fossil collections remain the staple diet for researchers. They act as the database that can be searched when new questions of old specimens need to be answered. Therefore collections should be cared for. Without taxonomy and associated rules of nomenclature it would be an impossible task to sift through the palaeontological literature and understand what fossils were being described. Many thousands of fossils have been named to date, and researchers from across the globe that meet can converse with each other by

Figure 3 An Eocene buccinid gastropod showing traces of the original shell coloration.

the study of fossils, and this section outlines our current understanding of the Earth's past environments and climates as revealed by ancient organisms, and documents how geologists have used fossils to develop an ever more precise geological timescale. Graptolites have been used to subdivide parts of the Lower Palaeozoic into short time slices of not much more than 4 million years' duration. How fossils are formed and the ways in which they are preserved is discussed, and a number of case studies of exceptional preservation in the fossil record are provided. Where are jellyfish or soft-bodied worms preserved? How often is original colour preserved? (Figure 3) Fossils have been the subject of study for several hundred years and over time their significance and the understanding of what they represented has changed from being objects of local curiosity to objects of huge scientific value. The second part of this book provides an outline of the major fossil groups, their form and function, and their links to living floras and faunas, and this will enable the reader to broadly identify and understand a broad range of invertebrate and vertebrate fossils. From plants, to microfossils such as foraminifera and radiolarians, through the myriad of invertebrate animals such as bryozoans, corals, molluscs and graptolites, to the vertebrates, this part also contains a synopsis of our own human origins, as well as a section on trace fossils. Readers will gain an overall appreciation of the evolutionary record of life on Earth.

A glossary of terms that are highlighted in **bold** throughout the text is given at the end of the book, which is illustrated throughout by photographs, line drawings and diagrams.

means of the taxonomic monikers applied to each fossil even though they may not understand each other enough to order a cup of coffee in each other's native language. Many answers about the nature of the past history of the Earth have been provided through

1.2 A chancy business: the preservation of fossils

Fossils are rare objects: for every one fossil that is preserved, many thousands of once living organisms have disappeared. Why should this be? The potential for preservation is controlled by various factors. If the organism consists of only soft tissue then the chances of its preservation in the fossil record are very slim indeed. In some exceptional circumstances soft tissues are preserved (*see* Section 1.8). When an organism dies, almost immediately the soft tissues begin to decay under microbial action, and if this is not arrested soon only the hard parts will remain. These hard parts such as shells and bone are therefore most commonly preserved. Depending on the circumstances, delicate skeletal parts may get damaged, disintegrate and be lost, whereas in general, more robust portions survive. Equally the mineralogy of the skeleton or shelly material may determine whether preservation happens. Commonly shells are composed of the minerals **calcite** or **aragonite** ($CaCO_3$), **silica** (SiO_2) as in radiolarians and many sponges, **apatite** (calcium phosphate) as in the case of many teeth or conodonts. Aragonite is more unstable than calcite and is prone to dissolving away faster.

A great deal of information about the living organism can be lost following its death and before it is fossilised, and the longer this takes, the greater amount of information loss will occur. Palaeontologists attempt to determine the nature of the living organisms by gaining an understanding of the geological and biological processes that have affected the organism after death and before it finally became fossilised. This is called **taphonomy** and can be divided into two stages: **biostratinomy** which covers death to burial, and **diagenesis** which is the period from burial to fossilisation. If one takes a living organism or assemblage of organisms, then on death information is lost due to soft tissue decay that leaves the skeletal hard parts. These can then get disarticulated, fragmented by scavenging organisms, attacked by microorganisms that **bioerode** the hard parts, abraded by knocking against other hard surfaces, dissolved or dispersed by water currents or occasionally wind. Once buried, the hard parts that remain can dissolve away, may be replaced chemically, or can get flattened and distorted, and finally the fossils may form.

Therefore even though a great number of fossils have been found and described it is not surprising that palaeontologists consider that the fossil record is incomplete. As we have seen, the chances of survival in the fossil record depend on many variables. Preservation may occur in one geological setting at a certain place on the planet at a particular time in the past, but may for a variety of reasons not occur in the same or similar setting at a different place and time. Shells and skeletons may be selectively removed from one environmental or geological setting due to the nature of their mineralogy or their morphological form but may survive in pristine condition in another setting. The comparison of living assemblages and those of the past can sometimes indicate those organisms that may have disappeared from the former (should they have been there in the first place). If this can be determined then important information can be gathered on the taphonomic processes that affected the **fossil assemblage** that is now seen. Palaeontology is all about data gathering, however thin and slight, and interpretation of that data, and it is worthwhile remembering that fossilisation is a chancy business.

There are a number of states in which a plant or animals may be fossilised:

Moulds and Casts. When a shell has been buried in sediment (or a pair of conjoined shells buried and infilled with sediment) that subsequently hardens and becomes **lithified** the shell may dissolve away leaving a **mould** (or a hollow where the shell had been). The mould of the external surface of the shell is called an external mould (Figure 4A) and of the internal surface an internal mould (Figure 4B). Should the mould become infilled a **cast** of the original shell shape is produced, which could be either an external cast of the external surface of the shell (Figure 4C) or an internal

cast of the internal surface of the shell. When only moulds are available for study palaeontologists will often produce a cast in latex rubber or plasticine for examination and photographing.

Original material. Sometimes the original material remains intact and chemically unaltered (Figure 5A). Many Cenozoic shells are preserved in this way, as are Woolly Mammoth carcasses frozen in ice, or Shark's Teeth from the Miocene (Figure 82). When preserved in limestone many graptolites retain their original **collagen periderm** and three-dimensional shape.

Permineralisation. Although appearing to be solid many shells are porous and also contain organic matter. Decay of the organic matter increases the porosity of the shells or bones, and these cavities may become infilled with calcite or silica or other additional minerals. This process is called **permineralisation** (Figure 5B). Delicate anatomical structures of plants such as **xylem** and **phloem** vessels and external **stomata** are preserved in this way.

Replacement. The skeletal material is replaced by another mineral such as iron pyrites or silica. Many fossils, such as the trilobites from Beecher's Trilobite Bed in New York, in

Figure 4 A. External mould of a single bivalve cockle shell; B. Internal mould of a conjoined pair of cockle shells; C. Exterior cast of a conjoined pair of cockle shells.

A   B   C

Figure 5 A–J Modes of preservation in fossils. A. Unaltered shell in the extant brachiopod *Magellania flavescens* from New South Wales, Australia; B. Permineralised shell of the brachiopod *Carneithyris carnea* from the Cretaceous of Norwich, England; C. Recrystallised shell of the brachiopod *Dielasma hastatum* from the Mississippian of Castleisland, Co. Kerry, Ireland; D. *Neuropteris*, a Pennsylvanian fern preserved in carbon as a compaction; ⊃

↻ **Figure 5 (cont)** E. *Merocanites*, a pyritised goniatite from the Pennsylvanian of Rush, Co. Dublin, Ireland; F. *Gryphaea* from the Lower Jurassic of England preserved in silica forming a concentric circular pattern indicative of the variety beekite; G. Internal mould of a turritellid gastropod; H. Flattened ammonite preserved in clay; I. A flattened graptolite *Monograptus* preserved in clay from the Silurian of Co. Dublin, Ireland; ↻

which appendages are preserved, or some late Mississippian goniatites (Figure 5E), which are replicated in iron pyrites (FeS_2), suggest that they were deposited either in deep water or in **anoxic** iron sulphide rich sediments. Chemical replacement of the original shell subsequently took place. Replacement by silica, which in some instances can be **opaline** (Figure 5F), can often faithfully replicate the surface details

of pressure from overlying sediments to a film of carbon. These are called **compressions** and appear as black shiny fossils (Figure 5D). This mode of preservation is most often seen in the Pennsylvanian Coal Measures plants.

Recrystallisation. Under pressure of burial under sediment the original skeletal material of shells may become **recrystallised** (Figure 5C). If this happens, the original internal layering and orientation of the crystallites of the shell will be lost. Many Mississippian brachiopods are preserved in this way and the shell is preserved in coarse calcite.

Flattening and distortion. Fossils may become flattened under the pressure of overlying layers of sediment. Ammonites which are hollow often become compressed in clays (Figure 5H). Flattened graptolites are preserved in shale where the original periderm is frequently replaced with **chlorite** or other **clay minerals** (Figure 5I). Although flattened, many of the **crustaceans** found in the Solnhofen Limestone have retained their original **chitin** skeletons. In the Middle Cambrian Burgess Shale, a diverse assemblage of weird and wonderful animals have been preserved in a flattened state as a thin film of clay. When fossils are subjected to **tectonic activity** they can become distorted. Many Cambrian trilobites exhibit such features (Figure 5J), and by using computer restorations the degree of strain and the direction of stress can be evaluated and can provide structural geologists with essential data for tectonic reconstructions.

⊃ **Figure 5 (cont)** J. Deformed Cambrian trilobite preserved in shale.

of the fossil, but internal skeletal features are often obliterated. The useful feature of **siliceous** fossils is that they can be easily extracted from the surrounding matrix using weak acids such as acetic acid or hydrochloric acid. Such extraction should only be attempted in a dedicated laboratory fitted with a fume cupboard.

Carbonisation. Plant tissues that were rich in **cellulose** may become altered under the effects

1.3 From the field to the laboratory: how to collect, curate and study fossils

It is probably true to say that the best collectors of fossils are firstly those with infinite patience, and secondly those who have seen a large number of fossils in the field. The first major collections were assembled in the late 1700s by various learned societies in European capitals, and in the 1800s by members of the landed gentry and by the more locally-based institutions of science and literature that developed in the growing urban centres across Europe. Today collecting is not the sole preserve of learned societies, nor indeed of academic palaeontologists, but the assembly of a good fossil collection is within the capability of almost anyone with a keen interest in the subject. However, if you are intending to collect fossils please give heed to a number of important rules that are outlined in the next section (1.4 Code of Conduct for Collectors).

Think before you collect – why are you collecting material? Is there a scientific question that needs investigation, and if so, what size of collection is required? If conducting a taxonomic revision of a particular **taxon** then a scientist may only need a dozen specimens, whereas a **palaeoecological** study that examines the interaction of organisms on a particular **bedding plane** or through a **succession** of bedding planes may need more material. Collect only the actual number of specimens that are needed. Sometimes the study can be conducted in the field without

resorting to collecting specimens – perhaps it might be possible to produce replicas of bedding planes for further study in the laboratory. In some cases it is necessary to collect specimens before they are lost to natural erosion, or lost due to construction of roads or buildings. In other cases a fossil has been located that represents a new taxon and the palaeontologists require further information about the geological context in which it lived. This would require them to visit the site from which the specimen had been collected so that they could gather data about the fossiliferous rocks and the nature and preservation of the fossils themselves.

Preparations for collecting

Many people enjoy building up collections of fossils – they are beautiful objects that tell a fascinating story both about the animals and plants that they represent but also about the history of our planet when these organisms were alive. In order to ensure that your collection survives and that it is scientifically useful there are a number of steps that a collector should take.

Assemble the correct equipment for collecting (Figure 6). You will need a geological hammer – that with a chisel-end is preferable to that with a pick-end; a set of cold chisels; a pair of goggles with wrap-around sides; a hard hat; a compass clinometer for measuring

Figure 6 Equipment needed for collecting and transporting fossils in the field.

strike and dip of the fossiliferous rocks; plastic self-seal or zip-lock bags of different dimensions for packing your specimens into; small clear plastic tubes for storing small specimens; newspaper and kitchen paper for wrapping specimens; pencils for note taking; indelible markers for labelling the specimen or bag in which it has been placed; a strong canvas bag or rucksack to carry your specimens from the collecting locality to your transport. It is also essential that you carry a hand-lens with a magnification of at least × 10, as this is useful for examining fossils in the field.

It is useful to learn something about the geology of the area you intend visiting. The Geologists' Association has published many guides on the geology of the UK, which are good sources of information. In order to record information in the field, you should also obtain a set of Ordnance Survey Maps for the locality so that you can mark on them the precise location where you obtained your fossils. A GPS is also useful in this regard. In tandem with these maps a set of geological maps can be beneficial. Information about the fossils and geology of the site should be recorded in a weather-proof file notebook. Spiral-bound notebooks are useless, as they fall apart in bad weather – you should invest in a purpose geological notebook such as one from the 'Rite-in-the-Rain' range. You should ultimately be able to cross-reference the data in the notebook with the specimens, so it is very useful to develop an easily remembered code for particular collections. For example, 2010/1/001 would refer to the first location visited and collected on the first collecting trip of 2010.

Collecting

You are now ready to start collecting. Ideally any trip should be undertaken with a friend or friends. If venturing out alone, notify someone of where you plan to visit and when you plan to return. If you don't return, then a search party can be sent out. When you reach the fossiliferous locality spend some time just examining it (Figure 7A). Work out where the fossils occur: are they scattered throughout the rock or are they concentrated in one place; do all the **beds** contain the same fossils? This will enable you to decide where is best to collect from. If the location is pristine it would be a serious mistake to hammer extensively at it. Look around the site and you may locate an equally fossiliferous but scruffy **horizon** from which it would be preferable to collect.

At many sites the fossils have been eroded out of **limestones** and **clays** and can simply be picked up. It is also a good idea to collect some of the **muds** or clays, as these can be sieved in the laboratory or at home later in the hope of finding smaller microfossils. At other places such as beaches, fossils can be found in loose stones that make up the shingle. If you have to extract a fossil using your hammer and chisel, first assess whether it is going to be difficult to obtain the fossil in this way without breaking or smashing it. If so, leave it where it is. Otherwise, carefully use the hammer and chisel to work around the fossil from some distance and eventually it should loosen. In well-bedded fossiliferous rocks use your chisel and hammer to split the rocks along bedding planes (Figure 7B). Often fossils will be revealed in this way on both slabs – these are known as the **part and counterpart** and should both be saved, as part of the fossil may be present on one slab and the rest remaining on the other slab.

Once a fossil has been collected it should be carefully wrapped in kitchen paper or newspaper, which should then be labelled with the location code and placed in a self-seal or zip-lock plastic bag. Record the find in the notebook and place the bag beside your rucksack. At the end of the collecting session

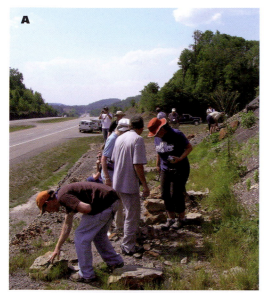

Figure 7 A. Collecting fossils in the Ordovician of Ohio, USA; B. Splitting fine-grained Jurassic limestone along bedding planes, Solnhofen, Germany.

carefully pack the specimens, now in their individual bags, into the rucksack, making sure that the large heavy samples are at the bottom and the more delicate specimens are at the top. If necessary, pack newspaper between the bags to stop them bumping against each other, which might cause damage during transport home. If you have collected material abroad it is tempting to mail it home rather than pay any excess baggage charges levied by an airline. Pay the excess charges, because that way you know your specimens will get home, and you will not have to worry about the vagaries of the postal system.

Curation

Once in the laboratory or at home unpack the specimens carefully, retaining the collection codes that go with each specimen. Fossils may need to be cleaned, and for most this can be done either by dry brushing off excess dirt, or when wet, under a tap. In some instances placing a specimen in an **ultrasonic tank** of **deionised water** for 30 seconds or less can remove adherents that won't be removed by brushing. Adding a **surfactant agent** can remove even more dirt. If you are unsure of what effect cleaning methods might have on your specimens, then clean one and note the effect before proceeding to another method. Some specimens may need to be trimmed or excess rock removed. This can often be done with a small chisel and small hammer, or engraving tool, but patience is required. Very delicate preparation is best done on a specimen viewed under a microscope using dental tools.

Large **macrofossils** should be labelled, and the best and permanent method is to paint a small rectangle in white gloss paint on the reverse of the specimen, let it dry, number the specimen in Indian ink, let it dry, and seal the number with a smear of clear varnish. The number that you use does not necessarily have to be the collection code but could simply be an **accession number** that will be recorded in the collection catalogue. The fossil should then be placed into an acid-free card box with a clear lid into which should be placed a tray label (Figure 8A). This should always carry the taxonomic name of the fossil, the name of the collector and date when collected, and the geological horizon and geographical location from where the specimen was obtained. If necessary place acid-free tissue around the specimen to stop it rolling around, without which it may be liable to get damaged. **Micro-fossils** should be placed in small sealed receptacles such as cavity slides (Figure 8B) which are available from many scientific suppliers, and the accession number and taxonomic and geological information should be inked onto the card sleeve.

Having curated the specimen the collector should list it in a Collection Catalogue that should replicate the information provided on the tray label. It is also useful to record in your catalogue where you have stored the specimen. Many collectors will use purpose-built specimen cabinets made of steel, or cabinets designed to hold 24 trays of microfossil cavity slides. Cabinets should be kept in a dry environment, as should any pyritised fossils. If stored in humid, damp surroundings the **iron pyrites** may become altered during a process called '**pyrite disease**', expand and produce sulphuric acid (Figure 8C). This destroys the specimens, labels and the trays in which they are kept. Storing your collection is a damp basement will lead to its rapid deterioration.

Figure 8 A. Specimen of *Euomphalus pugilis*, a Mississippian gastropod from Co. Kildare, Ireland, that has been well curated with information given on the accompanying tray label; B. Recent foraminiferan in a cavity slide; C. An ammonite that is suffering from pyrite disease which has caused the pyrite to expand and sulphur to exude from the surface.

Expert advice on the **curation** of geological collections can be obtained from the Geological Curators' Group which is an organisation affiliated with the Geological Society of London (www.geocurator.org).

Studying your fossils

The way that you study your fossil collection all depends on what information you require and where your interests lie. Some collectors only study particular taxonomic groups such as sharks' teeth or ammonites, and to aid them they have obtained relevant books, monographs and scientific papers (These are described in more detail in the following section 1.6). Others are interested in the interactions between groups, while other collectors are interested in fossils from particular geological horizons such as the Chalk or the Gault Clay. If you are interested in studying certain taxonomic groups only, then contact similar-minded palaeontologists who may have formed a formal society for just that purpose. Most professional palaeontologists, whether they be university or museum-based, will be delighted to discuss the significance of your collections and may suggest and help with publication of significant finds. Some collectors are teachers and use their specimens in the classroom, while others are members of local natural history or geological societies where they can display and talk about their collections and thus educate like-minded people.

For general information on fossils it would be worthwhile joining one or all of the palaeontological organisations. In the UK and wider afield the Palaeontological Association has been active since the 1960s and it publishes the journal *Palaeontology* six times a year, while the Paleontological Society is based in the USA and also publishes a journal.

Nowadays palaeontologists utilise a range of high-powered equipment to aid the study of fossils, and examination under **Scanning Electron Microscopes**, or analysis of skeletons and shells in a **Mass Spectrometer** is commonplace. For the general collector a standard **binocular microscope** is useful and not beyond most budgets, and on some of these a camera can be mounted for those interested in photographing their collections. Some fossils are best seen in **thin-section**, which most university departments routinely prepare, and for examination of these a **petrological microscope** is needed.

Over time some collections can be very extensive, running into hundreds if not thousands of specimens. Remember that the life of your collection will undoubtedly be longer than your own, and that you should make provision for its well-being after you have died. In many instances relatives don't show the same enthusiasm for your passion, and where this is clearly the case you should arrange that your collection will be deposited in a museum or institution where it will continue to be enhanced and cared for.

Collecting fossils can bring enormous pleasure and is most rewarding. It enhances our understanding of our planet's history. Collecting and curation takes time and effort, but this is adequately paid back both in the enjoyment and the scientific value that your collection can bring.

1.4 Code of conduct for fossil collectors

When you are collecting fossils please remember to adhere to a few rules. Many fossil sites worldwide are now protected by local laws and it is *illegal* to remove any geological specimens from them. In the UK many fossil localities are within National Nature Reserves, or have been designated **Sites of Special Scientific Interest** (SSSI) (Figure 9) or **Regionally Important Geological and Geomorphological Sites** (RIGS). In the Republic of Ireland a number of geological sites fall within Special Areas of Conservation (SAC) that are protected under European legislation, or are National Heritage Areas (NHA) under Irish law. The networks of Geoparks in Europe are protected, while in the USA it is illegal to collect on Federal-owned land. Find out which sites are protected and avoid using your geological hammer at them. Elsewhere use restraint when hammering and collecting.

It may also be illegal to remove material from one jurisdiction to another even if you legitimately purchased the material, and other material may fall under the **CITES legislation**. For many countries a Code of Conduct for fossil collectors exists – that for Scotland is very informative (Figure 10) and can be applied to most collecting situations. This can be downloaded from the Scottish Natural Heritage website (www.snh.org.uk) and should be read by all collectors.

In particular the following should be adhered to when collecting:

- Always ask written permission of landowners to visit and collect on their land. If given permission to collect, respect the land, close gates and do not damage fences and property.
- Do not enter working quarries. Quarries of any type whether they are active or disused can be very dangerous.
- Wear protective goggles and a hard hat when collecting.
- Do not climb cliffs, or hammer at the base of them, as this may cause rockfalls or landslides and lead to death or injury.
- Think before you collect. Do you require the fossils in the first place, and what will you do with them once collected?
- Do not over collect even if you think that there is an unlimited supply of fossils available to you. There certainly won't be, and such irresponsible collecting will lead to complete stripping of a fossil locality. There is no point in eventually assembling a large collection of miscellaneous fossils that end up filling a garage or attic.
- Record geological and stratigraphical information for your specimens.
- If you find anything unusual please contact your nearest geological museum for help identifying your specimen. If your material turns out to be of scientific importance, donate it to an accredited museum.

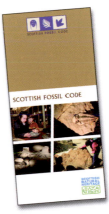

Figure 9 A. The Wren's Nest, Dudley, West Midlands, England – this is both an SSSI and a National Nature Reserve rich in Silurian fossil assemblages (© Natural England); B. Charmouth Beach, Dorset, England – this is part of the West Dorset Coast SSSI. The rocks in the cliffs are largely Lower Jurassic in age (© Natural England).

Figure 10 Leaflet on the Scottish Fossil Code.

1.5 Taxonomy: how to name and classify fossils

Taxonomy is the science of naming organisms and arranging them in a series of groups. The largest grouping is that of **Kingdom**, which is followed in descending order by **Phylum**, **Class**, **Order**, **Family**, **Genus**, **Species**, and **Subspecies**. Taxonomy is said to have begun on 1 January 1758, which is the date of publication of the 10th edition of Carl Linnaeus's *Systema Naturae*. He was a Swedish naturalist who realised that a consistent method of naming organisms was required because otherwise a plant given a name in Britain might be identical to a plant given a different name in Sweden, which would be terribly confusing.

Most fossils are simply known by their genus and species names (a **binomial** name), i.e. *Homo neanderthalensis* or *Tyrannosaurus rex*, and when printed these are italicised. The first letter of the genus name is always capitalised and is followed where known by the species name; if used frequently in a document it may be abbreviated to *T. rex* after first appearing. The name of the author who first named the fossil will also be given: i.e. *Vincularia megastoma* M'Coy, 1844. If a subsequent author placed the fossil into a different genus because they considered the original designation to be incorrect, the species name stays the same but the name would now be cited as *Baculopora megastoma* (M'Coy, 1844). The rules for plants are slightly different and both the author of the original name and the authors of the subsequent change would be cited.

Usually fossils are not given common names, unlike most living plants and animals, but occasionally such names exist as a hangover from local folklore, as in the Devil's Toenail (*Gryphaea arcuata*).

The names used for animals in palaeontology are governed by rules laid down by the International Commission on Zoological Nomenclature (www.iczn.org) which was established in 1895. Fossil plant names are governed by the International Commission on Botanical Nomenclature. If there are problems with nomenclature, then the relevant commission will decide on the correct solution.

When naming a new species or animal – and over 15,000 new species are named every year – a scientist has to publish a description in a valid journal or book. They also should cite and illustrate (if possible) the actual specimen on which the species name was based. This specimen is the **holotype**, and any other specimens in the collection are known as **paratypes**. These should be placed in a museum so that they are available to future researchers. The author of a name has the honour of deciding on the name, which utilises Latin or Greek roots. Often a fossil is named after an important person (*Caryophyllia smithii*, named after Sidney Smith, an expert on fossil corals, or *Arietites bucklandi*, named after the Rev. William Buckland (Figure 11)), a place (these names often end in –*ensis*), or after some characteristic feature that the fossil exhibits (*Cnemidopyge bisecta*, an Ordovician trilobite with a central ridge on its **glabella** (Figure 23)).

In palaeontology, of course, it can be difficult to prove if a species is a true biological species, although recent advances using DNA studies on fossils are beginning to yield some results and may be useful for species recognition in young fossils. Most species are still differentiated on the basis of morphological differences between fossils, which can be recognised through detailed measurements taken on them. These are **morphospecies**. In some cases binomial names were applied to different disarticulated parts of organisms, such as the Pennsylvanian plant *Lepidodendron*, to which names have been applied to its leaves, trunk, cones and roots; or to the different elements of conodonts. These are known as **Form taxa**. In trace fossil studies binomial names, too, have been applied to traces even though the maker may be unknown. *Oldhamia* from the Cambrian is an example of one such **ichnogenus** and *O. radiata* is an **ichnospecies** contained within that ichnogenus (Figure 12).

Resources for taxonomists

If you wish to identify fossils there are a number of useful resources available to you. The serious collector should build up a resource library of identification guides (Figure 13). The least expensive are the various identification guides that can be purchased in any good bookstore. The Natural History Museum in London produced the wonderful three volume guide to fossils of Britain that may be available through secondhand dealers.

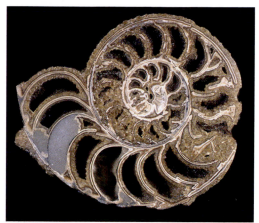

Figure 11 The ammonite *Arietites bucklandi* from the Lower Jurassic of Lyme Regis, Dorset, England.

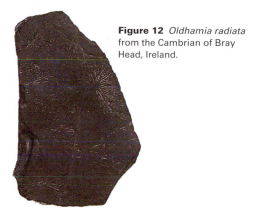

Figure 12 *Oldhamia radiata* from the Cambrian of Bray Head, Ireland.

Since the 1850s the Palaeontographical Society has been publishing monographs on the palaeontology of Britain. Each of these monographs provides detailed descriptions and illustrations of a particular fossil group of a certain age. For example, Thomas Davidson published a multipart monograph on the brachiopods (Figure 63), while the Rev. George Whidborne published on the Devonian fossils of southwest England. The Palaeontological Association and Paleontological Society both publish journals that frequently contain taxonomic information, while the former also produces a Guide Series to fossils of certain geological periods.

The bible for taxonomic palaeontologists remains the multi-volume series *Treatise on Invertebrate Paleontology* which commenced publication in 1953 under the editorship of the energetic Raymond Moore. Updated volumes continue to be prepared and copies can now be purchased either as hard copy or online. Each individual volume covers a major taxonomic group such as the Goniatites or Rugose and Tabulate Corals, and every genus is illustrated and described.

Collectors may also find it useful to visit their local geology museum and simply compare their specimens with those on display.

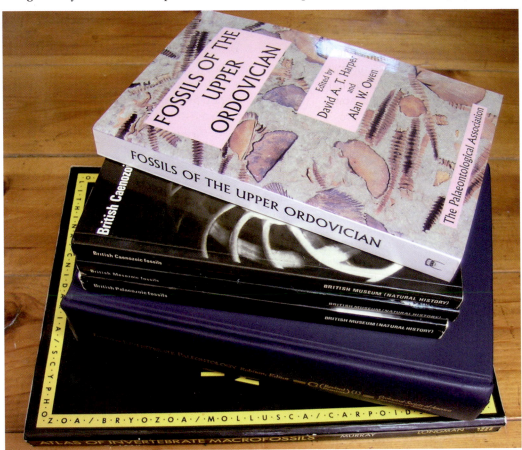

Figure 13 Various taxonomic books and monographs of use to palaeontologists. From top: *Fossils of the Upper Ordovician* (Palaeontological Association); *British Fossils* [3 volumes] (Natural History Museum); *Treatise on Invertebrate Paleontology* (Geological Society of America & University of Kansas) and *Atlas of Invertebrate Macrofossils* (Longman).

1.6 Uses of fossils

Scientists have long found the study of fossils to be of huge benefit in unravelling Earth history. In the infancy of palaeontology most studies were confined to taxonomy, in which many fossils were formally named. In the 1830s they were discovered to be useful for correlation, and many field geologists until the 1950s spent a great deal of time collecting material and documenting their finds as 'faunal lists' in publications. This work, while now considered to be somewhat routine, has provided palaeontologists with a rich database, and since the 1960s scientists have drawn on it to tackle other questions.

We now know a great deal more about the ages of fossiliferous rocks, thanks to biostratigraphical studies, and the rocks of a number of **Periods** have been correlated globally on the basis of their fossil content. We also have a better understanding of the past **palaeogeography** of our planet. Plants and animals display provincialism today and did so also in the past, and so the positions of former continents and oceans have been determined. Needless to say, geophysical investigations of the oceans and onshore have also provided clues to the past tectonic and continental plate movements.

Palaeontologists are now often interested in the biological evidence for lifestyles and any ecological interactions of the fossils that they study. Can any inferences about biological affinities, relationships, phylogenies and perhaps feeding habits be drawn out of hardpart morphology? Modern methodologies and equipment have helped reveal skeletal morphologies and mineralogies even in 3D. How did particular organisms interact with other organisms that they lived adjacent to in the past?

Palaeontologists are rather like genealogists as they try to find out about the **phylogeny** and ancestry of fossil groups and their evolution through time. Rather than browsing through old dusty tomes they carefully browse through the layers of fossiliferous rocks in the search for answers.

No doubt as new fossils come to light, and new methods are applied to pre-existing collections, palaeontologists will discover new information that will help assemble the complex history of life on Earth.

The sections that follow discuss some uses of fossils in more detail.

1.6.1 Palaeobiological history of life on Earth

The complexities of life on Earth and the numerous variations of the theme preserved in the rock record allow palaeontologists to assemble a huge database, the study and manipulation of which provides an inordinate number of questions and many answers. The progression of life is neatly documented, and the episodes where plants and animals have disappeared, and the reasons, may be deciphered too. The biological realm on Earth is constantly changing – species are evolving at the present but unfortunately are becoming extinct at the same time. Such developments in producing diversity have gone on ever since life first evolved on earth.

It is convenient to divide this section into four portions that correspond to the four major subdivisions called **Eons** into which the Earth's history is divided. The first three sections discuss the earliest life forms until the advent of the major shelly organisms at the beginning of the **Cambrian Period**.

Hadean history of the Earth

The earliest Eon on Earth is the **Hadean**, which began with its formation 4567 million years ago (**Ma**) (Figure 14) and continued until 4000 Ma. This was a time when the hot molten planet began to cool down, segregated into **core** and **mantle**, and small **proto-continents** of **crust** developed. Considerable **degassing** and volcanic activity produced an **atmosphere** that was rich in nitrogen, methane, carbon dioxide, ammonia, hydrogen sulphide and water vapour but lacked oxygen. Over time the protocontinents collided into each other and larger continental masses began to be accreted on the crust. The water eventually condensed and produced the oceans.

Archean history of life on Earth

Life began on Earth approximately 3500 Ma ago, some 350 Ma after the commencement of the **Archaean**, when a combination of gases, water, and the addition of an electrical spark instantly generated **amino-acids**. For a long time the mechanism of the genesis of life puzzled scientists until two Americans, Stanley Miller and Harold Urey, in 1952 conducted some experiments in a laboratory in Chicago.

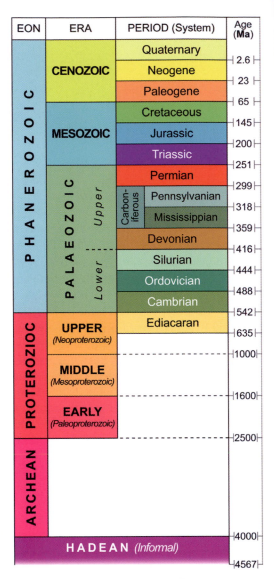

EON	ERA	PERIOD (System)	Age (Ma)
PHANEROZOIC	**CENOZOIC**	Quaternary	2.6
		Neogene	23
		Paleogene	65
	MESOZOIC	Cretaceous	145
		Jurassic	200
		Triassic	251
	PALAEOZOIC *Upper*	Permian	299
		Carboniferous Pennsylvanian	318
		Carboniferous Mississippian	359
		Devonian	416
	PALAEOZOIC *Lower*	Silurian	444
		Ordovician	488
		Cambrian	542
PROTEROZOIC	**UPPER** (Neoproterozoic)	Ediacaran	635
			1000
	MIDDLE (Mesoproterozoic)		1600
	EARLY (Paleoproterozoic)		2500
ARCHEAN			4000
HADEAN (Informal)			4567

Figure 14 The geological timescale. The colour codings for the various units are those prescribed by the Commission for the Geological Map of the World, Paris, France.

Figure 15 A. Fossil Stromatolites from the Cambrian Huqf Group, Huqf Desert, Sultanate of Oman. B. Modern Stromatolites from Shark Bay, Western Australia.

They simulated what they considered the conditions of the early Earth were like by filling a flask with methane and ammonia, caused a spark to cross a gap between some wires in the flask and later discovered proteins in a water reservoir at the bottom of the glass apparatus. The first life forms were tiny bacterial **prokaryotic** microbes that produced **stromatolites** found in the Oman and Namibia (Figure 15A), ancient forerunners of the modern examples in Western Australia (Figure 15B). These could photosynthesise and took carbon dioxide and water, and in the presence of sunlight produced sugars and oxygen. Slowly the levels of oxygen built up in the atmosphere.

Proterozoic history of life on Earth

The **Proterozoic** began 2.5 Ga ago when increased oxygen levels resulted in the precipitation of large volumes of iron dissolved in the oceans. This settled out and produced the **Banded Ironstone Formation** that is widespread in parts of Africa and Australia. By the middle of the Proterozoic (1.5 Ga) the atmosphere contained nearly 20% oxygen which is close to the proportion today, and **eukaryotic** cells first appeared. These have a **nucleus** surrounded by a membrane, and two types developed that gave rise to plants such as the algal **acritarchs**, and to animals (*see* Section 2.1). Continued photosynthesis caused the levels of carbon dioxide to plummet and the Earth was subjected to a number of glaciations around 700 to 650 Ma. Volcanic activity replenished the carbon dioxide to such an extent that limestone was precipitated extensively.

At 600 Ma a remarkable event took place in the evolution of life when **multicellular** forms began to be assembled. These **Ediacaran** floras and faunas were soft-bodied and most of this fossil biota is unrecognisable in terms of the modern biota. Frondose and disc-like organisms (Figure 16) abounded that may have had affinities with cnidarians, echinoderms and algae.

Phanerozoic history of life on Earth

By 540 Ma at the beginning of the **Phanerozoic** some animals discovered the secret of **mineralisation** and were able to extract dissolved

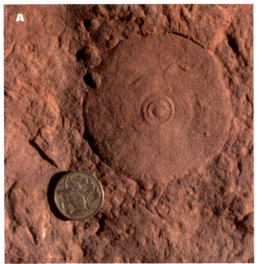

Figure 16 A. *Cyclomedusa plana*, an Ediacaran fossil from the Flinders Ranges, Australia; B–C. The oldest Ediacaran fossils dating from 603 Ma at Mistaken Point, Newfoundland, Canada.

Figure 17 *Cloudina* from Blasskranz, Namibia. Dating from 542 Ma these are the oldest fossil shells.

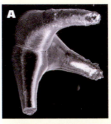

Figure 18 'Small Shelly Fossils': A. *Chancelloria*; B. *Halkieria*; C. *Lapworthella*; D. *Tannuolina*. From the Cambrian of Meishucun section, Maidiping, Sichuan, China dating from 542 Ma.

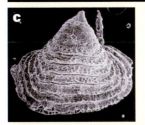

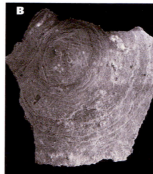

Figure 19
A. Precambrian–Cambrian boundary at Dvortsy Cliffs, Aldan River, eastern Siberia, Russia; B. Small Cambrian inarticulate brachiopod.

minerals from seawater and use them for the construction of hard shells. In one stroke the nature of the fossil record changed, and the potential for preservation had increased enormously. Organisms such as *Cloudina* from Namibia (Figure 17) are amongst the first shelly fossils, and this period was also marked by a proliferation at the base of the **Cambrian** of what palaeontologists group together as 'Small Shelly Fossils' or SSFs (Figure 18) whose affinities remain ambiguous. Use of the Scanning Electron Microscope has revolutionised the study of such material. The start of the Cambrian is marked by the appearance of these shells as well as others such as inarticulate brachiopods (Figure 19) which diversified rapidly and produced a significant fossil record.

The Phanerozoic contains three **Eras** which are characterised by a distinctive fossil content and rich in a variety of different biological groups. These groups are discussed in detail in later sections. The **Palaeozoic** Era (Cambrian to **Permian**) is dominated by invertebrates, animals without backbones (Figure 20). The first vertebrates were **Ordovician** fish, which were followed by amphibians in the **Devonian**. Reptiles first made an

Figure 20 Silurian shelly assemblage from Dudley, West Midlands, England.

appearance in the **Mississippian** but became more abundant in the **Mesozoic** Era (**Triassic** to **Cretaceous**), but at its end nearly 50% of all life on Earth was wiped out after it was hit by a large **bolide**. **Angiosperms** first appeared in the Cretaceous and many were aided in their success by evolving reproduction strategies that depended on insects that **coevolved** to reap the benefits of this activity. The **Cenozoic** Era comprises the **Paleogene** and **Neogene** (formerly called the **Tertiary**), when mammals became the dominant group on land, and they evolved into many different types.

Throughout the Phanerozoic the invertebrate biota both on land and in the oceans waxed and waned, and some groups such as brachiopods predominated in shallow marine environments early on before being displaced by the bivalves from the Mesozoic onwards. Other organisms have managed to remain almost unaltered over long periods of time and have been referred to by the moniker **'living fossils'**. *Lingula*, the inarticulate brachiopod, first appeared in the Cambrian 580 Ma ago and the plant *Ginkgo* (Figure 21) is a relative newcomer but has been around nearly 300 Ma since the Permian.

The history and story of life on Earth is highly complex and has been likened to a tree, 'the tree of life'. In the early stages there were few stem groups, but as time passed new forms evolved, these branched out, and became more diverse.

Figure 21 *Ginkgo*, an example of a long-ranged 'living fossil'. A Mesozoic fossil is on the right and a leaf from a living tree on the left.

1.6.2 Evolution and extinctions

Evolution

Without the process of evolution there would be a low diversity record of life on Earth. Even before Charles Darwin and Alfred Russell Wallace, who published a joint paper in 1858 on natural selection, the driving force of evolutionary change had been considered by some scientists. Natural selection depends on the passing on of advantageous inherited traits to the next generation that make them better placed to live to reproductive age and to continue the lineage. Other scientists such as Jean-Baptiste Lamarck had recognised that animals appeared to change over time. Lamarck suggested that they became increasingly complex and that they had the ability to cause change for themselves by adapting to their environment. This didn't satisfy many scientists, and the question of what precisely caused evolution stumped most theorists. Darwin and Wallace identified the mechanism of evolution, and while their paper did not reach a wide audience, Darwin's subsequent book *The Origin of Species* published in 1859 did. It caused a storm of debate amongst scientists, some of whom argued that the Earth was just not old enough for evolution to have taken place. Later the enormous longevity of the Earth was recognised, and there is no problem with providing adequate time for evolutionary processes to take place.

Natural selection is the process by which species are generated, and this has yielded a huge diversity of morphologies. Examination of the fossil record clearly shows that the Palaeozoic biota was dominated by invertebrates that were later joined by primitive vertebrates. The Mesozoic faunas were reptile-dominated

and the Cenozoic is mammal rich. Palaeontologists have built up a large database of information on evolution and now recognise evolutionary lineages in almost all fossil groups including conodonts, brachiopods, corals and mankind. The level of confidence of the veracity of these lineages and the phylogeny of the groups depends on the quality and volume of material available. Recording fossils preserved bed by bed in a long sequence of sediments can reveal evolutionary changes.

Within the fossil record there are numerous examples of co-evolution, of which that of the angiosperms and the insects is perhaps the best documented (Figure 22).

Figure 22 Leaf of the angiosperm *Viburnum* showing considerable damage by insects. Dakota Sandstone, Cretaceous of Ellsworth County, Kansas, USA.

Change, it has been shown, takes place at different rates. Nils Eldredge and Stephen J. Gould in 1972 argued for the process of 'punctuated equilibrium' where stocks remained largely constant for long periods of time, but that evolutionary change took place episodically and rapidly in short bursts, before the systems settled down to intervening stasis. Others, including the English palaeontologist Peter Sheldon, showed that evolutionary change could take place gradually but rather slowly. In a ground-breaking piece of work he showed that the number of ribs in the pygidia of eight trilobite genera of the Ordovician of central Wales increased steadily over a period of three million years (Figure 23) and that there was no evidence of punctuated equilibrium in his study.

Extinctions

Plants and animals have continually disappeared throughout geological time, either as small numbers of taxa over time or in a series of cataclysmic events that saw the removal of a considerable proportion of the biota. In many cases whole groups were wiped out and no modern examples are extant – there are no graptolites, trilobites or dinosaurs. In other cases most representatives in a group were removed but a small number of taxa survived, and from these new lineages evolved. The modern scleractinian corals that are major reef-builders are thought to have been derived from a single or small number of Palaeozoic corals. The ammonites that were highly successful in the late Mesozoic radiated from a single genus *Phylloceras* (Figure 55C) after

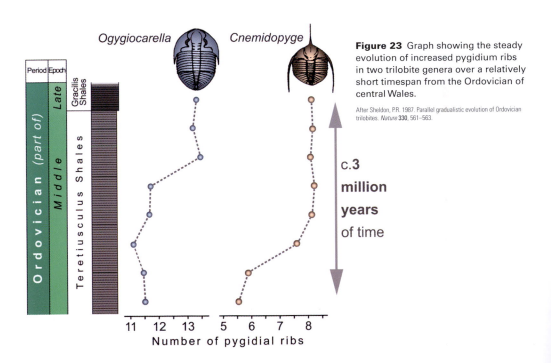

Figure 23 Graph showing the steady evolution of increased pygidium ribs in two trilobite genera over a relatively short timespan from the Ordovician of central Wales.

After Sheldon, P.R. 1987. Parallel gradualistic evolution of Ordovician trilobites. *Nature* 330, 561–563.

c.3 million years of time

Ogygiocarella   *Cnemidopyge*

Number of pygidial ribs
11   12   13   5   6   7   8

Period | Epoch

Ordovician (part of)

Late | Gracilis Shales

Middle | Teretiusculus Shales

all others were removed by the end-Triassic event. However, subsequently they failed to survive the end-Cretaceous event that also claimed the dinosaurs.

There have been at least seven major extinction events, of which six have occurred in the last 500 million years (Figure 24). According to the American palaeontologist Norman Newell, anything between 26% and 52% of all families became extinct during one of these events.

Vendian: This event occurred approximately 600 Ma ago and was marked by the loss of many stromatolite species and other algal groups such as acritarchs, as well as many metazoans such as the frondose Ediacarans (Figure 16). Part of the reason for this loss has been attributed to the activities of burrowing organisms that churned up the sediment (bioturbation) and made it unsuitable for many metazoans.

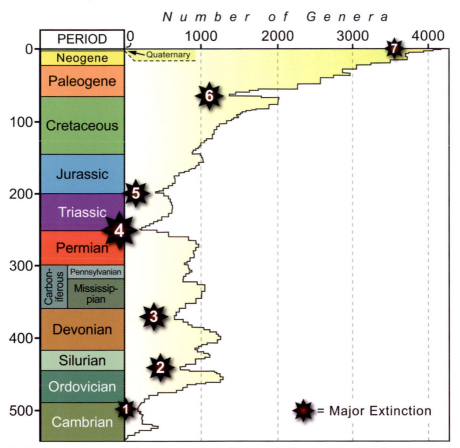

Figure 24 The seven major mass extinction events in the Phanerozoic. The size of the star indicates the severity of the extinction, while the graph shows the number of genera of marine organisms plotted through time. The numbers to left of the geological column refer to millions of years before the present.

Afer Rohde, R.A. and Muller, R.A. 2005. Cycles in fossil diversity. *Nature* **434**, 208–210.

Late-Cambrian: this, in terms of numbers of families lost, was the largest extinction with 52% going, but this may have been because the diversity of life was low at this time. Perhaps if the diversity had been greater, then more groups might have survived the changing circumstances. Gastropods, sponges and trilobites were major casualties.

End-Ordovician: Brachiopod, crinoid, echinoid and trilobite stocks were badly affected, with the trilobites in particular plummeting to only 20% surviving. A number of reasons for this loss have been promoted – a southern glaciation at the time would have affected the global temperatures, and allied to this, widespread ice formation would have led to a fall in sea level which might have exposed shallow-water marine shelf environments where many of the groups lived.

Late-Devonian: 30% of the biota was removed, including Brachiopods, Bryozoans, Cephalopods, Corals, Trilobites, and for the first time vertebrates – some Fish. These occurred at the end of the Frasnian Stage, when global temperatures fell, affecting coral reefs and many marine fish.

End-Permian: This was the first extinction event that resulted in the extinction of some major groups of life. Just over half of the families disappeared with ammonites, bryozoans and reptiles all being affected badly. Rugose corals, some crinoid forms, productid brachiopods and foraminiferans disappeared completely. Many reasons have been championed as being the cause of this major extinction event, but of these, the coming together of continental plates to form the super-continent Pangaea resulted in the loss of large areas of shallow-water shelf environments due to falling sea level and the ultimate loss of whole ocean basins, which would have had a catastrophic effect on the biota.

End-Triassic: Many ammonites, brachiopods, reptiles, Fish as well as all the conodonts were wiped out. Climatic changes on Earth at this time resulted in widespread arid conditions and changes to the vegetation which would have impacted on the terrestrial vertebrates. Falling sea levels or **regressions** would have affected the marine realm.

End-Cretaceous: This is certainly the most famous mass extinction event, as it was responsible for the demise of the dinosaurs. Apart from these animals the ammonites, rudist bivalves, and marine reptiles such as the pliosaurs and icthyosaurs became totally extinct. Many corals, echinoids and sponges also disappeared, but enough of them survived to mount a later recovery. The cause of this extinction event has been directly attributed to the collision of a large bolide with Earth. This hit the Yucatan Peninsula in what is now Mexico and resulted in large volumes of debris filling the skies. The dust blocked out the sunlight, plants died and were followed by the large terrestrial animals. The impact left a large crater at Chicxulub, which was later infilled with sediment, and was recently discovered using geophysical methods. Following the impact, acidic rain altered the pH of the oceans, and this had a profound effect on the marine stocks.

Pleistocene: During the Ice Age much of northern Europe and North America was affected by periods of extreme cold with the advance of glaciers and ice sheets oscillating with warmer interglacial periods. Today we are in one such interglacial, but the present lacks the large mammals that lived during earlier warm times. It has been estimated that most of the terrestrial mammals that weighed over 44 kg succumbed to extinction between 8000 and 18,000 years ago. In North America these mammals included sabre-toothed cats, mammoths and mastodons (Figure 98); in South America the large armadillos and sloths disappeared, while across Europe the Woolly Mammoth was lost, as was the Giant Irish Deer (Figure 97) that had roamed between Siberia and Ireland, and died out 11,000 years ago. It has been argued that the extinctions were caused by a change in climate from one with little seasonal variation to one marked by considerable seasonality. Another plausible cause was that humans, who had migrated to most of the globe by this time (Figure 100), hunted the large mammals into oblivion.

Evolution and extinctions are often closely linked. The sudden loss of many organisms from many environmental niches opens them up to opportunistic species and to those that evolve to suit the new conditions. Nevertheless, while mass extinctions can act as a driving force for evolution, and rapid change at that, it should be remembered that evolutionary change can also be very gradual. The Earth is a highly dynamic planet and many of its inhabitants have all too often been proven to be fatally prone to extinction when affected by change. This fragility has nevertheless had a beneficial spin-off for palaeontologists, as it has provided them with a rich and diverse biota for study.

1.6.3 Interpretation of ancient environments and climates

Today if one was to visit a tropical rainforest it would be a rather simple matter to conduct an ecological study and observe and record the plants and animals that you find there – simple, assuming that the recorder can identify the various taxa that are present. A similar exercise can be carried out in northern latitudes in the extensive coniferous forests of Scandinavia. That done, the two ecosystems could be compared in both biological and physical terms. Are the flora and faunas unique to each district, or are there some cosmopolitan species present? What are the climatic patterns and do they vary throughout the seasons? Where are the plants and animals living in each setting and what biological interactions can be deduced? These are difficult questions to answer and it would take a long period of continual assessment and study before patterns began to emerge.

Imagine, then, the difficulties presented to the palaeontologist who asks the same questions, but this time when standing on a bedding plane of Lower Ordovician limestones in Estonia, or three months later when examining another but younger Devonian limestone bedding plane in Ohio (Figure 25A). Gathering of data in palaeontology can often take more than a number of years, and scientists are fortunate that information is available for analysis that dates from the 1830s and even earlier. Ultimately it is possible to derive a great deal of information about the ecological habits of the organisms and even reconstruct a snapshot of what the environment may have looked like at particular times in the past (Figure 25B). Such reconstructions are often seen in museums

Figure 25 A. Bedding plane in Devonian limestone dominated by corals at the Falls of the Ohio, USA. B. Reconstruction of the Devonian assemblage (FOSP).

alongside the primary sources of the data – the fossils.

Palaeoecology is a similar discipline to ecology, which is the study of the inter-relationships of organisms (such as encrusting bryozoans hitch-hiking on trilobites (Figure 26)) and their environment, but differs from it in that information is lost through the passage of time and because many elements of the original biota may not have been preserved. Clues to the ecological parameters that affected organisms now found as fossils can be gathered from various sources: from **sedimentological** features such as bedding characteristics, the sediment type in the rock and the rate of sedimentation, from evidence of sorting and transportation, fragmentation and accumulation of shells and skeletons. Fossils can be **autochthonous**, being preserved *in situ* where they lived, or they may be **allochthonous**, having been transported to where they ultimately became buried. Where large numbers of identical shells are accumulated together in a **coquina** (Figure 27) this shows that they have been moved by water currents. If highly fragmented, they probably came a long way, but if pristine, transport distance would have been negligible. Were the deposits **marine** or non-marine? Were they deposited under high energy or low energy conditions or on open shelf settings like the **continental shelf** of western Europe? Perhaps the deposits indicate a lagoonal or mangrove setting seen today in the Persian Gulf or Florida, or rather a deepwater **abyssal** setting similar to those on the modern Mid-Atlantic Ridge. The Mazon Creek fossils were entombed in a large estuary similar to that of the Mississippi that **progrades** into the Gulf of Mexico. Many of these questions can be answered through a

Figure 26 Trilobite with epizoan bryozoans encrusting its cephalon (CMNHS).

Figure 27 Coquina composed of thousands of shells of the extant bivalve *Fragum hamelini* from Shark Bay, Western Australia.

close examination of bed geometries and the arrangement of fossils in the rock.

The study of the ecology of a single taxon is called **autoecology** and data can be gathered about the size of specimens, the mineralogy of the shell, the morphology of the animal or plant, and their ecological preferences in terms of salinity, light and water temperature. *Harpes* and *Cryptolithus* are distinctive fringed trilobites that were blind, which suggests that they had no need of eyes and would have lived beyond the **photic zone**. The fringes may have carried sense organs and also may have been a structural adaptation to benthic living on muddy substrates by spreading out the weight of the **carapace**.

Ecological study of **communities** is called **synecology**, and here palaeontologists can determine much of what they gathered through synecological studies. Individual taxa in a community can occupy various niches where they prefer to live. Some are **infaunal** or **epifaunal**, **pelagic** or **benthic**, **free-living** or **sessile**. Equally thay can be classified on the basis of their feeding habits. Plants **photosynthesise**, producing food, and are called **producers**. They are eaten by primary **consumers** (who represent the lowest **trophic level**) and as these are eaten by ever larger secondary and tertiary consumers (at higher trophic levels) energy is transferred through a food chain, or if the pattern is complex, through a foodweb that links all the organisms in a community. Corals, bryozoans, sponges, graptolites and many bivalves and brachiopods are **suspension feeders** that trap food suspended in the water. Many crustaceans, echinoids, gastropods and annelid worms are **deposit feeders** that browse and scrape food from the sediment surface, and **predators** such as many crabs, cephalopods, gastropods and dinosaurs actively seek their food. By careful study these complex feeding patterns can be determined. Generally when a **niche** becomes available, for example through rapid sedimentation covering and killing animals on the seabed, the new available space is colonised by a low diversity **pioneer community**. Over time these become covered by sediment and joined by a greater diversity of groups, eventually producing a **climax community** where

the palaeoenvironment cannot support any more diversity or numbers. A bed by bed study of the fossils can reveal these changes in the community dynamics over time.

Additionally, information about species **diversity** and species distribution on beds can yield important information about the configurations of the continents and oceanic basins of the past. A classic study of the fossil brachiopods of Wales showed that during Silurian times a distinctive zonal arrangement could be seen. This was interpreted as reflecting different water depths that were favoured by particular taxa. *Lingula* lived closest to the shore, whereas *Clorinda* lived adjacent to the **continental slope** in deeper water (Figure 28). We can determine the dynamics and age of fossil populations by measuring the sizes of individual shells – if they are all equisized, then it is a probability that only adults are present in the record and that the **juveniles** have been transported away, or that they migrated as larvae and settled far from their parents. During the Ordovician, trilobites exhibited **provincialism**; those found in northern Scotland are distinct from those of a similar age from Wales. This suggests that when alive they inhabited a wide ocean that must have drifted together over time. The distribution of different fossil groups has allowed palaeontologists to build up a record of the past positions of continents and oceans, and these palaeogeographical studies have been augmented in the last half century by geophysical data.

In recent years new techniques have been developed in palaeontology that have utilised instruments such as the mass spectrometer. This allows geologists to measure the composition of small amounts of shell or skeleton, and in the last two decades this has been

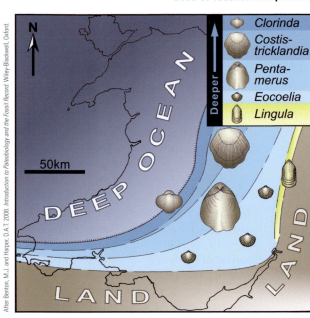

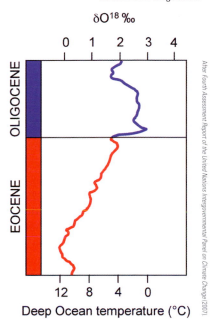

Figure 28 Map showing the distribution of Silurian brachiopods across the Welsh Basin. The assemblages reflect deepening water to the west.

Figure 29 Graph showing the change in surface water temperatures during the Eocene and Oligocene.

important in climate studies. Oxygen removed from seawater by organisms is trapped in shells as they are secreted. Two **stable isotopes** of the element commonly occur – O^{16} and O^{18} (another is rather rare), and depending on the water temperature, these isotopes are precipitated in different ratios. In cold water O^{16} is preferentially evaporated, and so shells are enriched in O^{18}. Following measurement of the ratios in a mass spectrometer and comparison to a standard measure, which was determined from a belemnite guard, the original seawater temperature can be worked out. This can perhaps tell us something of the latitudinal distribution of many pelagic and benthic species such as foraminifera, and reveal whether they were polar, tropical or equatorial in the past. Some corals and bryozoans recorded isotopic variations that were regarded as indicating winter and summer seasonal temperatures,

and this allowed the age of the skeleton to be determined. Palaeontologists have analysed many fossils derived from boreholes drilled through ocean sediments by the Deep Sea Drilling Project. Analysis of **microfossil** O^{18} isotope values during the Eocene and Oligocene have demonstrated that the temperature slowly dropped during the Eocene and then plummeted at the Oligocene, shortly after the commencement of which there was a **glaciation** (Figure 29). These studies may have implications for modern-day and future climatic estimations.

Palaeontology is a science that allows its practitioners to re-evaluate the past environments and conditions in which the fossils lived, and the assembly of pieces to the jigsaw of past life remains as much a valuable occupation today as it did in the 1830s.

1.6.4 Biostratigraphy and the concept of zone fossils

Biostratigraphy is the use of fossils in determining the relative age of rocks. This helps provide an understanding of the historical setting in respect to the succession in which that unit is situated, and fundamentally biostratigraphy allows palaeontologists to correlate units from locality to locality, whether they be close to each other across a valley or more widely spaced across large continents or ocean basins.

The origins of biostratigraphy go back to the work of the English surveyor William Smith (1769–1839) who spent a great deal of his time also collecting geological material encountered during his surveys. This led to his publishing in 1815 the first large-scale geological map of England and Wales. Critically, he recognised that particular horizons or even individual strata could be characterised through identification of their constituent fossils, and that their relative ages could be determined. This was enshrined in 1819 in his 'Law of Strata Identified by Fossils', and once this was generally accepted, geologists had a means whereby they could correlate with confidence strata from one area to another district over a considerable distance.

Smith's scheme, however, was rather coarse, and the units identified by him were thick and therefore of a long duration. Studies in the Jurassic rocks in Germany by Friedrich Quenstedt and continued by Albert Oppel demonstrated that the succession could be divided into rather short units, now called **biozones**, by means of ammonites. Oppel erected over 30 such zones identified on the basis of a very distinctive or especially abundant taxon. These distinctive fossils are known as **zone fossils** (or sometimes referred to as **index fossils**) and usually provide the name of the biozone. The *bucklandi* Biozone in the Sinemurian of the Lower Jurassic of England is based on the ammonite *Arietites bucklandi* (Figure 11) and spans a period of six million years (189.6–195.5 Ma). The younger *humphriesianum* Biozone in the Middle Jurassic of England is based on another ammonite *Stephanoceras humphriesianum* (Figure 55D).

To be a useful zone fossil it must have a short time range and be found in a variety of different sedimentary rocks laid down in different environments. They should be easy to identify from complete specimens or even fragments, and obviously for global correlation have a cosmopolitan distribution. Those with a more regional distribution that were confined to a single, but large marine basin, would be useful as zone fossils but for more restricted and local correlation.

Patterns of assemblages, occurrences, appearances and disappearances of fossils and other variations in their distribution have led to the recognition of different types of biozones. A Total-range biozone is defined by the range of one taxon only (Figure 30A) and a Concurrent-range biozone by the overlap of the ranges of several taxa (Figure 30B). A Partial-range biozone is distinguised by the disappearance and appearance of some taxa occurring within the range of another taxon

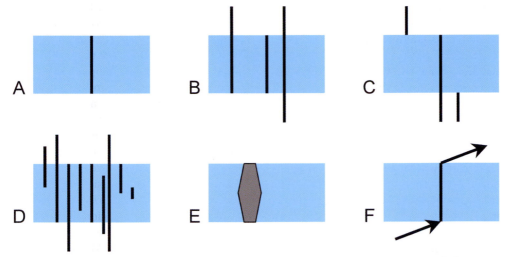

Figure 30 The major types of Biozones. A. Total-range biozone defined by the range of one taxon; B. Concurrent-range biozone defined by the overlap of the ranges of several taxa; C. Partial-range biozone defined by the disappearance and appearance of some taxa occurring within the range of another taxon; D. Assemblage biozone defined by the ranges of nine taxa; E. Acme biozone defined by the range of a dominant or abundant taxon; F. Consecutive-range biozone defined by the range of a taxon within an evolutionary lineage.

After Benton, M.J. and Harper, D.A.T. 2008. *Introduction to Paleobiology and the Fossil Record.* Wiley-Blackwell, Oxford.

(Figure 30C) and an Assemblage biozone contains many taxa whose ranges delineate upper and lower boundaries to the biozone (Figure 30D). In some instances a geological unit can be dominated by one fossil taxon whose range defines an Acme biozone defined by the range of a dominant or abundant taxon (Figure 30E). Within evolutionary lineages it may be possible to determine the range of one taxon on either side of the taxon from which it evolved and the later taxon which evolved from it. This provides a Consecutive-range biozone (Figure 30F).

Over geological history different fossil groups have proved to be useful zone fossils for different geological periods. In the Cambrian, trilobites and their associated crawling trace *Cruziana* have been used successfully in Scandinavia. Graptolites are especially useful in correlating Ordovician and Silurian rocks. *Didymograptus bifidus* the 'tuning fork' graptolite (Figure 31) is an instantly identifiable fossil that is confined to the Arenig Stage of the Lower Ordovician.

Palynomorphs – spores, seeds and pollen of terrestrial plants – have been successfully utilised to erect detailed biozones for successions spanning the late Devonian high into the Pennsylvanian. These microfossils are found in terrestrial rocks such as coal, but were also washed into marine settings and are rather abundant in deepwater marine shales, which can often yield assemblages of over 100 species. Their distribution has enabled correlation of non-marine and marine rocks. The stratigraphical distribution of *Cyrtospora cristifera* and other palynomorphs (Figure 32A) has allowed **palynologists**, who study

Figure 31 *Didymograptus bifidus*, the 'tuning fork' graptolite, which is a useful zone fossil indicative of the Arenig stage of the Lower Ordovician.

this material, to correlate with precision rocks in Missouri in North America with those in Ireland (Figure 32B).

In the Mesozoic, zones of durations of approximately one million years in the Jurassic and Cretaceous have been erected on the basis of ammonites and belemnites, a precision that in terms of geological precision is short and remarkable. Within the Cenozoic, foraminifera have provided biostratigraphical precision that has proved important in the search for evidence of palaeo-climate change.

In the last thirty years or so palaeontologists and stratigraphers have expended a great deal of energy formulating local and global correlation schemes. In the 1970s the Geological Society of London published a series of booklets for each geological period

in which details of local successions were given, together with illustrations that showed the correlation of units as small as Formations from district to district across Britain and Ireland. More recent schemes have been more ambitious, having successfully erected global standards in terms of biostratigraphic units for different geological periods, and now the boundaries between many important subdivisions of the geological timescale are well defined and marked by hypothetical 'golden spikes' at designated boundary sites. Much of this work was carried out under the direction of the International Union of Geological Sciences of UNESCO, which had established sub-commissions for each geological period. The work of the IUGS-sponsored International Geological Correlation Programme (IGCP)

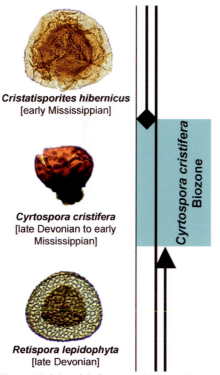

Cristatisporites hibernicus
[early Mississippian]

Cyrtospora cristifera
[late Devonian to early
Mississippian]

Retispora lepidophyta
[late Devonian]

Figure 32 (above) A. Spore partial-range biozone
defined on the presence of *Cyrtospora cristifera*
throughout this part of the late Devonian to early
Mississippian sequence, the disappearance of *Retispora
lepidophyta* after the first third of the sequence and
the appearance of *Cristatisporites hibernicus* in the
final third of the sequence; (below) B. Correlation
of *Cyrtospora cristifera* biozone rocks at Hannibal,
Missouri, USA (left) with those at the Old Head of
Kinsale, Co. Cork, Ireland (right).

drew together geologists from different countries to tackle pertinent research problems, and it continues to promote biostratigraphical studies, amongst others, to the present day. One of its recent projects, number 491, focused on the Middle Palaeozoic vertebrates.

Biostratigraphy is one of the essential disciplines in geology, and underpinning it are the fossils that have provided the means and basis for the relative dating of rocks and stratigraphical successions. In many parts of the globe studies need to be advanced so that a fuller biostratigraphical framework can be achieved, and this can only be done through the study of fossils.

1.6.5 Development of the geological timescale

Stratigraphic geology concerns the layering of the Earth, and over some 200 years geologists recognised that certain sequences were distinctive, and these packages were given particular names to identify them. This gave rise to a nomenclature that meant it was easy to correspond with other scientists, all of whom understood the meaning and relative positions of different packages of rocks. **Stratigraphy** is one of the fundamental disciplines of geology, and the use of fossils in making some sense of

stratigraphy remains important. Stratigraphic protocols are now controlled by the International Union of Geological Sciences, which has a number of committees or commissions to oversee particular geological periods/systems or boundaries between them.

The discipline of stratigraphical study is split into three sub-branches: **lithostratigraphy**, **biostratigraphy**, and **chronostratigraphy**. Lithostratigraphy is concerned with the naming, defining and description of rock units in terms of their physical characteristics. In biostratigraphy, as we have already seen, fossils define the relative ages of rock units and aid correlation. Chronostratigraphy concerns itself with absolute time that has been computed through measuring the decay rates of radioactive elements found in igneous bodies such as batholiths or dykes, which often cross-cut fossiliferous successions. If one can compute the absolute date for the injection of the body, then that sets the upper time limit for the rocks intruded into.

Today, much effort goes into defining internationally agreed boundaries between various units known as '**Systems**', '**Series**' and '**Stages**' whose absolute geological age is known. A 'System' comprises all the rocks between its lower and upper boundary, and it corresponds to the division of geological time known as a '**Period**' so that one can talk of the Permian System rocks that were deposited during the Permian Period.

The geological column (Figure 14) is divided into four major divisions or Eons – the earliest is called the Hadean, a term coined by the American stratigrapher Preston Cloud (1912–1991) after Hades, the Greek underworld. This part of Earth history is poorly known and extends from its initial formation to the formation of crustal rocks and the accretion of proto-continents. Next in order come the Archaean ('primitive') and the Proterozoic ('first life') and the **Phanerozoic**. The former is largely unfossiliferous, while life began to burgeon during the Proterozoic. The Phanerozoic ('visible life') comprises the last 600 million years or so when life on Earth was, and remains, diverse and abundant.

The Phanerozoic is divided into three Eras: the Palaeozoic (ancient life), Mesozoic (middle life) and Cenozoic (recent life). The Eras are further divided into Periods, which are shorter and usually, but not always, defined by distinctive rock types and recognised by their particular fossil constituents. Nearly all Periods are themselves sub-divided into lesser **Epochs**, and even into **Zones**, which can comprise a very short span of time, and which are recognised by diagnostic **zone fossils** or zone assemblages.

Many geological periods were named and defined in the 1830s, although some are earlier. The term '**Jurassic**' was first used in 1795 by Alexander von Humboldt, the noted scientist and traveller. The '**Cretaceous**' was named in 1822 by a Belgian geologist and contains the distinctive Chalk (a pure limestone) that is always associated with the Cliffs of Dover, but crops out elsewhere. In 1822 the Rev. W.D. Conybeare and the printer William Phillips named the '**Carboniferous**' for the marine limestones, overlying sandstones, and coal measures, successions seen in England and South Wales. Later this stratigraphic term was replaced in the USA by '**Mississippian**' in 1869 for the limestone, and '**Pennsylvanian**' in 1891 for the coal. Recently it was agreed that these American stratigraphic names should be used globally. In 1834 the German geologist

Friedrich von Alberti recognised a succession of three units, which include the New Red sandstone, which he called the 'Triassic'.

In the 1830s two great friends, Adam Sedgwick and Roderick Impey Murchison, began mapping the geology of Wales. Sedgwick named the northern portion '**Cambrian**', while Murchison gave the southern succession the name '**Silurian**'. Soon, however, Murchison began to encroach on Sedgwick's Cambrian, which led to a falling out of the friends, but not before 1839, when they named the '**Devonian**' for the tract of clastic rocks in north Devon. Only in 1879, after the death of the two men, was the situation clarified when Charles Lapworth, a Birmingham palaeontologist, inserted the **Ordovician** Period between the Cambrian and the Silurian. This was based on his study of graptolites.

Murchison, later in 1841, named the 'Permian' for rocks that he had seen during a visit to the Urals in Russia. Recently the terms 'Paleogene' and 'Neogene' have been introduced for the now obsolete period 'Tertiary', and these contain a number of subdivisions called Epochs, named by the English geologist Charles Lyell in 1833: **Eocene**, **Miocene** and **Pliocene**. Later he added the **Pleistocene**, which is now most associated with the last Ice Age and a characteristic fauna of large mammals including the Mastodons, Mammoths and the Giant Irish Deer.

The development of the geological column or geological timescale relied heavily on the description and recognition of fossils that remain one of the fundamental sources of primary information for geologists working today.

1.7 Fossil Lagerstätten: exceptional preservation of fossils

Lagerstätte is a German word that refers to any rock containing economically interesting or important components. Palaeontologists have modified the term as '**Fossil Lagerstätte**' (plural: -en) meaning any rock that contains fossils sufficiently well preserved to merit examination. Well-preserved fossils are not that common and are particularly useful because they demonstrate more completely the range of plants and animals in assemblages. Discussion of how fossils came to be preserved can be found in an earlier section (1.2). There are two categories of fossil Lagerstätten: **concentration** and **conservation**.

Concentration Lagerstätte, as the name implies, refers to exceptional concentrations of fossils. Organisms or their hardparts may accumulate in great numbers by a variety of mechanisms or in certain geological settings. Many modern mammals migrate seasonally, and those such as Wildebeest can drown in large numbers while crossing swollen rivers. Their bodies will be swept downstream and eventually their bones trapped in slower-flowing parts of the rivers. In a similar way, bone beds such as the Ludlow Bone Bed, or the younger Rhaetic Bone Bed (Figure 33), both of western England, may have been assembled in these placer sediments. Many fossils of small Paleogene mammals such as mice and other rodents have been found in crevices and fissures that form concentration traps in ancient **karstic** features;

others, in cave systems in limestone districts where these concentrations are known as **condensation deposits**. Shells can be transported after death and cluster together to form shelly banks. These, comprising mussels, are often seen on modern beaches at high-water level or above. If fossilised, these would form a coquina such as those of oysters in Portland Limestone from the Jurassic of England or the beautiful ammonite assemblage of Marston Magna in England, in which the original mother-of-pearl shell is preserved (Figure 1).

Conservation Lagerstätten may be poorer in terms of numbers of fossils preserved than in concentration Lagerstätten, but the quality of preservation is far greater. Soft and delicate tissues are retained, and these can provide enormous detail about the morphology of the organisms, and in many cases these fossil Lagerstätten are the only examples in the fossil record of certain taxa. In order to preserve soft tissue, one must arrest or stop decay after death altogether. This can be achieved in a number of ways: freezing, salting, lowering the **pH**, drying them out, or removing oxygen. Many thousands of Woolly Mammoth have been found in Siberia in ice, and recently attempts have been made to extract viable sperm for possible crossing with modern elephants. In Germany during the Jurassic, poorly circulating lagoonal waters had elevated salt concentrations, and in northeast

Figure 33 Rhaetic bone bed with rounded pieces of black bones, coprolites and scales from the Upper Triassic of Aust Cliff, Gloucestershire, England.

Spain a Woolly Rhinoceros has been preserved in a salty oil shale. If animal carcasses fall into tar shales such as those at Rancho La Brea in California, or oil shales as at Grube Messel, or if they sink into bituminous shale deposits such as found at Holtzmaden they can be preserved in stagnant conditions. In some situations echinoids and crinoids may be preserved complete, which is rare, thanks to their being buried rapidly in sediment following a storm. The Mississippian crinoid faunas of Crawfordsville, Indiana and Hook Head, Ireland are such **obrution** deposits.

Until the 1960s the number of fossil Lagerstätten known could be counted on the fingers of one hand, but since then, palaeontologists have targeted particular geological horizons in which they believed that the right conditions for fine preservation had been present at the time of their formation. This, added to the development of equipment such as the Scanning Electron Microscope and advances in light microscopy and X-ray techniques,

have increased the number of fossil Lagerstätten known. In addition to these new finds are those fossil occurrences known since the 1800s, which have been re-evaluated using present-day techniques. Now over twenty fossil Lagerstätten have been described in detail, and annually new valuable information on the history of life derived from them and elsewhere comes to light.

The major fossil Lagerstätten are:

- The **Apex Chert** (Archaean) of Western Australia is 3500 Ma old. While not producing many fossils, they are certainly exceptional as they are the oldest microfossils known. These filamentous fossils have been compared, by some experts, to **cyanobacteria**, which if this comparison is correct shows that they were capable of photosynthesis.

- **Doushantuo Formation** (Precambrian) of China dates from approximately 600 Ma and is one of a number of Proterozoic Lagerstätten. Of particular significance was the recovery of microscopic multicellular phosphatised acritarchs and algae, and that of embryos of unknown animals in very early stages of development. Other examples of early Lagerstätten are from successions 1000 Ma old in Siberia, and 800 Ma in Spitsbergen, which have produced fossils of bacteria, algae, and other eukaryotic cells.

- **Ediacaran biota** (Ediacaran). In 1946 a suite of unusual fossils were found by geologist Reginald Sprigg in the Flinders Range of South Australia, preserved on the underside of sandstones. Various jellyfish and medusae were present (Figure 16A), as were other radial fossils such as *Dickinsonia* and *Tribrachidium*. Some scientists have argued that, while difficult to interpret, the Ediacaran biota may include representatives of early echinoids, molluscs and cnidarians. In 1958 in Leicestershire, England, similar fossils were found, of which the frond-like sea pen *Charnia* was the most impressive (Figure 34). Ediacaran fossils, as they are now known, have since been found in various other localities, including Newfoundland at Mistaken Point (Figure 16B–C), Namibia, Russia and China, and they provide a tantalising view of the biota prior to the onset of mineralised skeletons and shells. It has been suggested recently that microbial mats of blue-green algae may have facilitated the preservation of the soft-bodied organisms by overgrowing them when they lived on a soft muddy seabed. An influx of sand then produced casts or moulds. On account of these wonderful fossils from Australia and elsewhere, a new geological period was recently formally accepted. This is the Ediacaran Period, which comes immediately before the Cambrian.

- The **Chengjiang biota** (Middle to Lower Cambrian) was located in 1984 in China, and has produced over forty species of arthropods, twenty sponges, some brachiopods, and various worms. There are some similarities to other Cambrian faunas such as the Burgess Shale, but the Chinese Lagerstätte lacks echinoderms.
- The **Burgess Shale** (Middle Cambrian) was first discovered by Charles Doolittle Walcott of the United States Geological Survey in 1909 (Figure 35A). In the summers that followed, he and his family excavated a great deal of material from

Figure 34 *Charnia masoni*, an Ediacaran frondose organism from the Precambrian of the Charnwood Forest of Leicestershire, England.

a small quarry in between Mount Field and Mount Wapta in British Columbia in Canada (Figure 35B). In the 1970s and at the present time, material is being collected for scientific study. The Burgess Shale yielded phyla which it is

Figure 35 A. Charles Doolittle Walcott of the United States Geological Survey and later the Smithsonian Institution, Washington DC, discoverer of the Burgess Shale fossils. Following the find in 1909 he spent many summers excavating fossils with the assistance of his family. B. The Walcott Quarry in the 'Phyllopod Bed', located between Mount Field and Mount Wapta, British Columbia, Canada; C. Portion of *Anomalocaris canadensis*, the largest animal found in the Burgess Shale; D. *Ottoia prolifica*, a priapulid worm that lived in U-shaped burrows in the sediment.

thought lived in a shallow-water shelf area, but which were swept into a deep abyssal setting by **turbidity currents**. The **biota** is found in shale preserved largely articulated and with soft tissue present in a white clay mineral, and is dominated (43%) by unusual arthropods such as *Marella*, which resembles a trilobite, and *Anomalocaris canadensis* (Figure 35C) which was a major predator. Priapulid worms (Figure 35D), corals, molluscs, sponges (Figure 48A), brachiopods, echinoderms, the highly enigmatic organisms such as *Hallucigenia sparsa*, and an early chordate *Pikia gracilens* are also recorded. This biota demonstrates the nature of the major faunal **radiations** that metazoans underwent in the Cambrian, and it is interesting to note that many unusual **clades** are not found later. Perhaps this is a function of their poor preservation potential, or else that they simply became extinct later. Similar preservation to that of the Burgess Shales has been discovered in earlier successions on Peary Land in northern Greenland.

- Orsten Shales (Upper Cambrian) of southern Sweden have yielded small **ostracod** arthropods complete with cuticular tissue, eyes and limbs. The fossils are preserved in coaly limestone **concretions**, and were extracted by digesting the surrounding rock with acids.

- The Soom Shale (late Ordovician) crops out in a north–south trending sequence 50–75 km east of Capetown, South Africa. Here many arthopods including the unusual trilobite *Soomaspis splendida* occur, as well as a large conodont animal complete with preserved muscle fibres, brachiopods, ostracods and fine examples of **epizoan** brachiopods attached to **orthoconic nautiloids**.

- The Rhynie Chert (early Devonian) is an important and perhaps unique fossil Lagerstätte because it preserves examples of some of the earliest plants that colonised the land surface, as well as various animals such as spiders and mites associated with

them. At the time land plants were generally small, standing no more than a few centimetres tall, but they had developed vessels in their stems that could carry water and foodstuffs. *Rhynia*, which, as the name suggests, was first described from the chert, had slender branches, which when mature carried small olive-shaped sporangia on their tips. Other plants included *Asteroxylon*, whose upright stems developed from a series of underground rhizomes. The **Rhynie Chert** assemblage lived around a small hot spring near what is now the village of Rhynie in Aberdeenshire, Scotland. During the early Devonian the **hydrothermal** regime that produced the hot springs also resulted in the deposition of silica in the form of opal, and it is this that the fossils are preserved in. The delicate structures and animals are best seen in thin-section. Similar preservation of plant material is known in the Princeton Chert in British Columbia which is middle Eocene in age.

- The Hunsrück Slate (early Devonian) occurs adjacent to Koblenz, Germany. Extraction of these slates for roofing has gone on for several hundred years; it was only in the 1960s that it was discovered that they contain exceptional fossils. Examination of the slates under X-ray revealed beautiful material replicated in the sulphide iron pyrites. Starfish and brittlestars are completely articulated, and arthropods, including the unique *Mimetaster*, are complete with delicate limbs and mouth parts. It is thought that the organisms are well preserved because they were rapidly fossilised.

- The Mazon Creek (Pennsylvanian) of

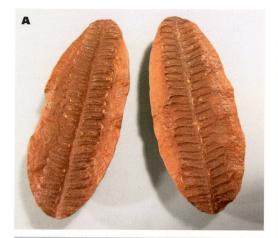

Figure 36 Fossils from the Pennsylvanian of the Mazon Creek of Illinois, USA. A. Fragment of a fern from the Braidwood assemblage; B. *Essexella,* a jellyfish from the Essex assemblage; C. *Tullimonstrum* gregarium (or the 'Tully Monster'), an enigmatic fossil from the Essex assemblage.

Illinois, USA biota is preserved in small **siderite** nodules that developed in the Francis Creek Shale Member that was associated with thick coal seams below. Strip mining removed the shales, which were left as spoil-heaps, and the nodules weathered out. When split, 3D fossils are found to be inside. The Mazon Creek can be divided into two assemblages. The non-marine component, called the Braidwood Assemblage, is rich in plants (Figure 36A), with lesser bivalves, fish, crustaceans, spiders, scorpions, leeches, and centipedes, while the estuarian component, called the Essex Assemblage, contains plants, jellyfish such as *Essexella* (Figure 36B), shrimps, insects, fish, the enigmatic *Tullimonstrum gregarium* (Figure 36C) (which is the State fossil for Illinois) and other elements. These were preserved as fragments washed into rivers and buried. As the organic matter decayed, gasses were produced that enveloped the organism, and this pocket was converted to siderite by bacteria.

- The Posidonienschiefer (Jurassic) of Holtzmaden in Germany consists of thick black **bituminous marls** and bituminous limestones. The marls were deposited under stagnant conditions over a long period of time, while the limestones and concretions contain the well-preserved fossils, complete in many instances with soft-tissue. Some ichthyosaurs and plesiosaurs are surrounded by a black film which represents the outline of the body. The deposit also contained the spectacular crocodile *Steneosaurus* (Figure 93), fish, sharks, belemnites, and crinoids attached to floating logs. A major question concerning ichthyosaur biology was answered in 1892 when a specimen was discovered that showed that the kink always seen in the dorsal portion of their backbone was natural and not an artefact of preservation, rather it supported the large lower lobe of the dorsal fin.

- Solnhofen Limestone (Jurassic) is a fine-grained limestone called Plattenkalk that is often thinly bedded and easily split. Formerly engraved for use in lithography, it is now utilised largely for paving. Sandwiched in an area north of Munich between the Rivers Danube and Altmühl are a number of quarries that have long exploited this stone (Figure 37A). In 1860 a find of a single feather caused a sensation in the scientific world, which was mesmerised by the discovery the following year of a fossilised bird, which was named *Archaeopteryx lithographica* (Figure 37B), and subsequently purchased by the British Museum (Natural History) in London. Since then the lithographic limestones have yielded many thousands of fossils, some of which have retained some soft tissue such as muscles. Fish dominate the biota (Figure 37C) that

also contains pterodactyls, horseshoe crabs, shrimps, dragonflies, squids, ammonites, stalkless crinoids, and rare algae. Preservation took place in stagnant lagoonal waters into which animals fell from the sky or were washed in during storms from open-marine settings in the Tethys Ocean to the south. Evidence of high salinity is seen in the fish and in the rare specimens of *Archaeopteryx*, where the drying out effect on the muscles of the backbone has resulted in their becoming significantly flexed.

- Grube Messel (Eocene) consists of an **oil shale**, nearly 200 m thick at its greatest, which was commercially exploited from the 1890s. This deposit is situated 30 km southeast of Frankfurt, Germany. The oil shale is actually a clay that contains 5–20% oil, and which was deposited in the bottom of a small lake under **anoxic** conditions during the Eocene 47 Ma ago. It is a stagnation deposit in which algae, fungi, pollen, freshwater sponges, fish, insects, turtles, birds, and 35 species of mammals, including **marsupials**, bats, horses and the recently described basal primate *Darwinius massillae* (nicknamed 'Ida') met their untimely end.

- **Baltic Amber** (Eocene) is derived from the resin of pine forests that grew in western Finland and eastern Sweden during the Paleogene 35 Ma ago. Amber was originally deposited in sediment called blue earth in northern Poland. From this it was reworked and redeposited as secondary or allochthonous deposits in eastern England (Figure 38), western Denmark, and along the northern coast of Germany. Fossilised within the amber are numerous

Figure 37 A. Quarry in Jurassic Solnhofen Limestone, Germany; B. The early bird *Archaeopteryx lithographica*; C. *Coccoderma*, a bony fish from the Jurassic Solnhofen Limestone, Germany.

Figure 38 Eocene Baltic amber carved into the shape of a fish. This piece was collected near Yarmouth on the east coast of England.

flies, spiders (and their webs), beetles, and millipedes trapped within the gooey resin. Following equinoctial storms, amber hunters can be found combing the beaches along the Baltic looking for this beautiful material. Long prized for jewellery, today a buyer needs to take care that they are not buying modern insects entombed in coloured resin. Amber dating from between 40 and 15 Ma found in the Dominican Republic also contains an abundant insect assemblage, and older Cretaceous amber is known from Africa and elsewhere.

● **Rancho La Brea** (**Pleistocene**) in Los Angeles contains a diverse assemblage of animals that roamed this part of North America during the last Ice Age. Naturally occurring pools of **asphalt** trapped over fifty species of mammals that were unfortunate to stumble into the area (Figure 39). These mammals range in size from mice to wolves, to sabre-tooth cats, to huge mastodons and mammoths, but amphibians, reptiles, birds, fish, and invertebrates including insects have also been recovered. These **tar-pits**, as they became known, were exploited for their asphalt, particularly in the late 1800s when it was used in road construction. Today a number of the worked out pits have been flooded and are contained in a public park where reconstructions of some of the fossilised mammals can be seen. Altogether many thousands of bones were removed for study, and excavation on the site still continues for palaeontological research.

Figure 39 Rancho La Brea Tar Pits, Hancock Park, Los Angeles, California, USA. A flooded asphalt pit with methane bubbling through the water, and life-sized models of struggling and onlooking mammoths.

1.8 Early ideas on the nature and significance of fossils

Fossils have long been a source of fascination for mankind. Four thousand years ago Neolithic man in western Ireland placed Mississippian brachiopods and cephalopods inside a passage grave, and later fossil sea urchins were found surrounding a skeleton interred in a later burial in Dorset, England. While not understanding the scientific and biological significance of such fossils, these early civilisations must have deemed them to hold some spiritual meaning, particularly the coiled nautiloids whose shape is a frequent motif in their decoration and carvings.

In Greek and Roman times fossils were thought by some philosophers to be organic in origin. Leonardo da Vinci also understood fossils to be the remains of marine animals, and even though the Italian forms that he was familiar with occurred at high elevations, he realised that meant that the strata in which they were found had once been under water.

In medieval times certain fossils gained local importance, and folklore concerning their origins was perpetuated. St Cuthbert was associated with the monastery at Lindisfarne off the coast of Northumberland, and the disarticulated stem ossicles of crinoids found in the bedrock were thought from 1671 to be rosary beads and called 'St Cuthbert's Beads'. In North America crinoids were called Indian Beads, and Native Americans strung the Cambrian trilobite *Elrathia kingii* for necklaces.

The first attempt in Britain to describe fossils was made in 1677 by Robert Plot, of the Ashmolean Museum, Oxford. In his *Natural History of Oxfordshire* he illustrated some tiny petrified horse's heads, and called them Hippocephaloides. These were known locally as ''Osses 'Eds' but are in fact the internal moulds of the bivalve *Myophorella incurva* from the Upper Jurassic (Figure 40A). Nine years later Plot illustrated 'Screwstones', which turned out to be crinoid stems from the Mississippian of Derbyshire.

Other examples of fossil folklore include the 'Devil's Toe Nails' which are actually the Lower Lias (Jurassic) bivalve *Gryphaea arcuata* (Figure 40B). These distinctive fossils also feature on the coat of arms of the town of Scunthorpe in England. Captain James Cook hailed from Whitby in Yorkshire, and no doubt he was familiar with 'Whitby Snakestones' (Figure 40C). It is possible that he knew them to be *Dactylioceras commune* or *Hildoceras*, Upper Lias (Jurassic) ammonites, specimens of which locals carved snakes' heads from the last body chamber. Many locals believed them to be petrified snakes banished by St Hilda, who died in 680, and they also feature in the coat of arms of the town and in some local trading tokens dating from the eighteenth century. To many people of Leicestershire, the bullet-shaped guards of belemnites were left behind at the spot where lightning bolts struck the Earth.

A

Figure 40 A. ''Osses 'Ed ' – the internal mould of the Upper Jurassic bivalve *Myophorella incurva* from the Portland Oolite of the Isle of Portland, Dorset, England. The ears are the infill of the beak and the eyes the adductor muscle scar; B. 'Devil's Toe Nail' – the Jurassic bivalve *Gryphaea arcuata*; C. 'Whitby Snakestone' – *Dactylioceras commune*, a carved ammonite from the Jurassic of Whitby, Yorkshire, England; D. Edward Lhuyd's fossil 'flat fish'; ⊃

C

B

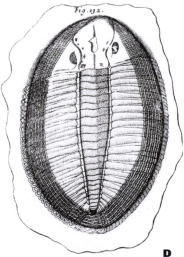

D

⊃ **Figure 40 (cont)** E. Faked carved 'fossils' in the form of slugs and Hebrew characters described by Johann Beringer in 1726.

In Europe the first scientifically rigorous attempt to understand fossils was that of Conrad Gestner of Zurich (1516–1565) who published *De Omni Rerum Fossilium* (A Book on Fossil Objects, Chiefly Stones and Gems, their Shapes and Appearances) in 1565, which included the first published images of fossils. He recognised the link between modern-day living crabs and their fossil counterparts. For many this link was difficult to comprehend, and nearly two hundred and fifty years were to pass before fossils were generally accepted as being the remains of animals and plants. One academic who understood this link was the teacher Johann Beringer of Germany, but

he was the victim of an elaborate hoax. Two of his colleagues at the University of Würzburg carved 'fossils' of birds, slugs, eggs, comets, symbols and even spiders' webs and left them scattered throughout the countryside for Beringer to find (Figure 40E). Unfortunately for him, he published a detailed monograph on the biota in 1726, and once he discovered the hoax he attempted, unsuccessfully, to retrieve all the 'fossils' and copies of his book. Edward Lhuyd (1660–1709) Keeper of the Ashmolean Museum, Oxford, understood fossils to be the remains of organic organisms, and he published a famous volume *Lithophylacii Britannici Ichnographia* in 1699 in which he illustrated some fossils including dinosaur teeth, and also 'flat-fish' from the rocks of central Wales (Figure 40D). The latter are the trilobite *Ogygiocarella debuchii*.

However, it was only in the late 1700s that the nature of fossils and their possible biological affinities began to be more fully understood. This was due, in no small measure, to the development of anatomy and medicine. Baron Georges Cuvier (1769–1832) of Paris was an anatomist who examined fossil mammals from the Paleogene of the Paris Basin and recognised the close links between some fossil forms and living species. He also identified proof for the concept of extinction, noting that mammoth were extinct relatives of modern elephants. At much the same time in Britain the popularity of fossil collecting and palaeontology was accelerated through the finds made of marine reptiles on the south coast by Mary Anning (1799–1847). In Oxford in 1824 the Rev. William Buckland (1784–1856) produced the first systematic description of a fossil reptile, which he named *Megalosaurus*, and the following year Gideon Mantell named *Iguanodon* based on a tooth and some bones found by his wife in 1822 at Tilbury. Nearly two decades later in 1841 the London-based anatomist Richard Owen coined the term 'Dinosaur', which was rapidly adopted. These ferocious animals became well known to the general public in Britain following the Great Exhibition of 1851, when large life-size models of them and other fossil animals were built by Benjamin Hawkins, a sculptor, and erected at the Crystal Palace in Sydenham, southwest of London.

By the 1800s palaeontology had become a subject for serious research and was included in many university curricula for the first time. Specialist monographs were published, and thanks to the work of William Smith geologists began to understand that fossils could be used for biostratigraphy. Fossils played a vital role in the development of the geological column as we now know it (*see* Section 1.6.5) (Figure 14). Charles Darwin (1809–1882) published his *Origin of Species by Means of Natural Selection* in 1859 and partially drew on his early researches in geology in devising his ideas on transmutation or **evolution**.

Since the 1950s palaeontologists have become less concerned with taxonomy and spend more time researching palaeoecology, the interactions and dynamics between fossil organisms, investigating the biological meaning of fossils and how they lived, and unravelling their ancestry or **phylogeny** through evolutionary studies. Modern techniques are now providing details of skeletal and shell structure and how it might have been precipitated, and the advent of DNA testing is beginning to throw up many intriguing questions. Research on stable isotopes of oxygen are ongoing and have indicated the nature of past climatic conditions, and those on nitrogen and carbon suggest what food types were eaten by former inhabitants of the Earth. In the last decade there has been an upsurge in research on Precambrian life, and our understanding of this formerly 'hidden' time in life history is now being illuminated. There is no doubt that the fossil record still has a great deal to reveal, and for that the fascination continues.

Part 2

Fossil groups

2.1 Algae and vascular plants

Algae and other lower plants

The earliest plants were algae, which probably first appeared on Earth over 3500 million years ago. As one might imagine, they have a poor fossil record, but a record that is nevertheless significant and important. Once life had originated, the first cells were prokaryotic, they had a cell membrane and a single strand of DNA bearing several thousand **chromosomes**. Among the early prokaryotics were blue-green algae (cyanobacteria) which have been found in the Apex Chert of Western Australia. Later examples, some 1500 Ma younger, are known from Canada, and take many shapes including filaments and spheres.

Blue-green algae were responsible for the construction of stromatolites. These are mound-like structures in which successive layers of sediment and blue-green algae build up. They were restricted to shallow-water environments where, importantly, they could photosynthesise. This process combined carbon dioxide and water to produce water, sugars and oxygen. As the Archaean progressed, extensive areas of stromatolites resulted in increased abundance of oxygen in the atmosphere. Modern-day stromatolites are only found in Western Australia, at Shark Bay north of Perth (Figure 15B). Fossil forms are known from many Precambrian successions.

Approximately 1000 Ma ago another form of cell evolved. This was the eukaryotic cell in which the DNA was confined within a nucleus surrounded by a nuclear membrane. Other organelles became an integral part of the cells – chloroplasts in the case of plant cells and mitochondria in the case of animal cells. It has been suggested that the incorporation of anaerobic bacteria produced mitochondria, and photosynthetic bacteria became chloroplasts. This was an endosymbiotic response by several prokaryotic cells which was self beneficial. Whatever the reasons and mechanisms for the development of prokaryotic cells, they were the building blocks of all higher forms of life.

While blue-green algae continued to thrive, new **unicellular** and **multicellular** forms such as red algae, green algae, brown algae, diatoms, and dinoflagellates evolved. Single-celled organisms that contain a cell nucleus are placed in the Kingdom **Protista**. Multicellular plants are placed in the Plant Kingdom. Many of the red, green and brown algae are familiar to us today as the seaweeds in the oceans or pondweeds in freshwater. These have a poor fossil record due to being largely composed of soft tissue, although *Peltoclados* from the Silurian of Ireland has been interpreted as possibly having charophyte (green algae) affinities. Rare Lower Jurassic examples of algae are known from the Solnhofen Limestone of Germany.

Diatoms are a group of algae that first appeared in the Jurassic. They are predominantly unicellular, but some multicellular forms are present in the phytoplankton. The tiny cells are enclosed in a split casing made of silica, and when they die these cases fall to the sea floor, where in deep waters they accumulate as diatomaceous ooze or **diatomite**. This material has been extracted commercially from Quaternary and Tertiary deposits in California and Denmark, and has various uses. Alfred Nobel discovered that if it was mixed with nitroglycerine, it made the highly explosive material easier to transport. The resultant mixture he called dynamite.

Dinoflagellates are unicellular algae with a large nucleus, a visible chromosomal apparatus with yellow or brown chloroplasts and two ventral flagellae. Some lack an external covering, and others have a cellulose coat or theca. Many species are part of the **plankton** of the surface waters of marine, lagoonal and lacustrine environments. In favourable conditions they may multiply rapidly and produce 'red tides', causing the water to become toxic.

Examination of the Cretaceous chalk under a microscope reveals that it is made up of tiny calcitic circular plates that resemble a Frisbee. These plates are **coccoliths** which form the outer test of a **coccolithophore**, unicellular calcareous plankton which could photosynthesise. When they died, the plankton sank slowly through the water column and the soft tissues decayed away, leaving the coccoliths which formed an ooze. During the Cretaceous thick deposits developed and lithified into the chalk.

Vascular plants

A major evolutionary step in the evolution of plants was the appearance of vessels that were capable of transporting water and dissolved foodstuffs. These vessels, called xylem and phloem in modern plants, evolved in the Silurian approximately 425 Ma ago, and gave rise to a number of tiny vascular plant genera. *Cooksonia*, named for Isabel Cookson, an Australian palaeobotanist, is the best known and had stems that ranged in length from 2 mm to 60 mm long, which when mature bore oval to spherical-shaped sporangia on their terminations (Figure 41). Other taxa included *Hostinella, Psilophyton, Gosslingia* and *Hollandophyton*. These vascular plants may have originated on algal mats that floated close to the shoreline, but that were well equipped to

Figure 41 *Cooksonia*, an early Silurian vascular plant with short stems and bulbous sporangia developed on their tips. Moneygall, Co. Tipperary, Ireland.

make the transition to a terrestrial environment. The Rhynie Chert of northeast Scotland is Devonian in age and gives an early insight into the greening of the planet. Plants have been preserved in chert, which has revealed very delicate structures. The complete assemblage points to vegetation growing around small lakes. All of these plants belong to the Rhyniophyta, which are included in the Subkingdom Tracheobionta which are vascular plants that produce spores.

Other more advanced Tracheobiontids included the Progymnospermophyta, which were capable of producing woody tissue but which could not produce seeds. They include the Devonian–Mississippian plant *Archaeopteris*, and may well be the ancestors of the modern seed plants.

The Devonian through the Mississippian and into the Pennsylvanian saw a major radiation of plant groups such as horsetails (Equisetophyta), club mosses (Lycophyta), true ferns (Pteridophyta), seed ferns and conifers. The Pennsylvanian swamp forests developed in Europe and North America in subtropical latitudes in a similar terrestrial–marine environment to that of the mangrove swamps presently found in Florida, although the type of vegetation differed greatly. Horsetails such as *Calamites* (the stem) (Figure 42A) and *Annularia* (the leaves) grew to 10 m in height while the club moss *Lepidodendron* attained tree-sized proportions, being 30–40 m in height. Complete plants have not been found, and palaeobotanists have given its various parts different 'generic' names: *Lepidodendron* (bark) (Figure 42B) *Stigmaria* (roots) (Figure 42C), *Lepidostrobus* (cones), *Lepidophyloides* (leaves). The true ferns were also plentiful and included typical Pennsylvanian coal

measures taxa such as *Psaronius* and *Pecopteris* (Figure 42D–E). When the plants died they became waterlogged and partially decayed. However, much plant material survived and when these thick deposits were compressed by overlying sediment, coal was produced. Much of the plant material ended up preserved in or above coal seams as carbonate compressions, and this coal is still extracted in Poland and western Pennsylvania in the USA. Extraction continued from several coalfields in Britain until the late 1970s. Fossil Grove in Victoria Park, Glasgow contains the stumps of eleven *Lepidodendron* trees. These were found in 1887 and shortly afterwards the site was covered with a pavilion which continues to protect this fossil forest from the elements, and in which the general public may view it. Another exceptional site where Pennsylvanian forests are preserved is Joggins Cliff, Nova Scotia in Canada, where nearly complete trees have been excavated.

Many of these plants produced **spores** (Figure 43) and pollen, part of their reproductive organs, and these have proved very useful in biostratigraphy and **palaeogeographical** studies (*see* Section 1.6.4). Known collectively as **palynomorphs**, they would have been dispersed by wind and other agents and now are found in a variety of terrestrial and marine sediments, but particularly in coals and shales. Research on Mississippian and Pennsylvanian examples has shown that the plant communities yielding the palynomorphs were highly diverse and comprised upwards of one hundred different species.

As many palynomorphs are covered by a hard outer layer, they tended to be preserved in sediments. Being organic, they have been used as indicators of the past temperatures

Figure 42 Pennsylvanian Coal Measures plants. A. The horsetail *Calamites*; B. *Lepidodendron*; C. *Stigmaria*; D. The leaves of true fern *Pecopteris arborescens*; E. Cross-section through the stem of *Pecopteris* from Newcastle-upon-Tyne, northeast England. A and C–D from England; B from Ohio, USA.

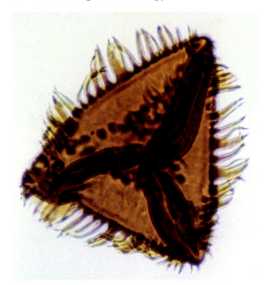

Figure 43 A plant miospore, *Diatomozonotriletes magnus* from the Scremerston Coal Group, Mississippian, East Lothian, Scotland.

(or thermal maturity) that their host rocks reached. These rocks may have been heated due to adjacent igneous intrusions or simply as a result of deep burial. When the spores and other organic matter heats up it darkens, and the degree of darkness can be calibrated to give a temperature reading or range of the '**maturity**' of the rock.

In the Mesozoic many of the older plant stocks dwindled (Figure 44B) and were replaced by the **gymnosperms** – the seed-bearing plants. There were several groups including the conifers, the glossopterids, the ginkgoales, and the cycads. During the Mesozoic Period a large continent was situated in the Southern Hemisphere. Called Gondwana, it comprised South America, Antarctica, Australia, Africa and India. Here a distinctive flora developed, now known as

the *Glossopteris* **flora** (Figure 44A). The flora is dominated by the groups mentioned above, and many botanists consider the landscape to have resembled pockets of vegetation dominated by cycads which are still living today in South Africa and New Zealand. The modern *Ginkgo biloba* has remained virtually unchanged for just over 200 million years.

In Arizona large numbers of Triassic conifers of the genus *Araucarioxylon* have been preserved almost intact in the area now known as the 'Petrified Forest National Park'. The Navajo Native Americans thought that the logs were the bones of the Great Giant Yietso. In some cases they used the logs as a building material. These fossils are related to the extant Monkey Puzzle Tree (*Araucaria*) and to the Wollemia Pine (*Wollemia nobilis*) which until recently was known only from fossils, the youngest being 2 Ma old. In 1994 a number of living plants were located in the Blue Mountains of New South Wales, Australia, and now the tree is available commercially from garden centres. *Wollemia* is a good example of a **Lazarus Taxon** which is one that has a long gap in its fossil record; it disappears and then reappears. This can be due to preservational or sampling bias.

Towards the end of the Aptian Stage in the late Early Cretaceous, the **angiosperms**, the flowering plants, first evolved, and they rapidly diversified throughout the period (Figure 44C–D). These are now the dominant plants on Earth today. The early success of the angiosperms is probably due to the presence of insects, which had evolved somewhat earlier. Success was founded on the ability to pollinate and therefore set seed, and in the insects the plants had found a useful mobile partner. As time passed then plants developed particular

Figure 44 A. Cycad from the Jurassic of India that comprised part of the *Glossopteris* flora;
B. Portion of the trunk of *Tietea singularis*, a Permian tree-fern from near to Araguaína, Tocantins State, Brazil;
C. *Magnolia* in full flower in April. Varieties of the plant group first evolved nearly 100 Ma ago; D. Maple is a
tree that has a fossil record extending back at least 40 Ma.

flower shapes and structures which could
only be pollinated by particular insect species.
Such reliance on each other is now common
in the modern plant world, but it does have
some inherent dangers should either party
become reduced in numbers.

2.2 Unicellular animals: Foraminifera and Radiolarians

Within the fossil record are preserved a number of different unicellular animals, and these make up an important component of what are termed microfossils. These are fossils that require a microscope to study, although a number of forms can develop quite large shells that are visible with the naked eye. Those unicellular organisms that possess a cell nucleus are grouped into the Kingdom Protista and include both plants (diatoms and coccolithophores) and animals (**foraminifera** and **radiolarians**)

Microfossils may be extremely abundant in certain rock types; for example, much of the limestone used to build the pyramids of Egypt is made up of a large foraminiferan belonging to the genus *Nummulites* (Figure 45). The Greek writer Strabo, who died in AD 25, described them as being similar to lentils, but which he noted were considered to be the remains of the food of the hardworked slaves who constructed the pyramids! Many Palaeozoic limestones contain small phosphatic tooth-like microfossils called conodonts.

Given their abundance, extraction of microfossils from their entombing rock can provide a large suite of specimens for research. Siliceous fossils contained in limestone can be removed by dissolving the surrounding rock with acids, while fossils in mudstones can be revealed using chemicals that break down or disaggregate the fine-grained sediments. All of these methods should only be carried out under laboratory conditions by experienced scientists.

The study of microfossils has proved particularly useful in oil and gas exploration where drilling has provided only a limited amount of material. Exploration companies will drill through many hundreds of metres of rocks using a drill bit attached to a long tube. The drill bit is armed with three or more rotating heads that break up the rock, and these small rock fragments are carried to the surface in a mixture of lubricating drilling mud. Once at the surface the samples can be removed to the laboratory, where thin sections can be made or microfossils extracted prior to study.

Amongst the common unicellular animals are foraminifera and radiolarians. Other microfossils include diatoms, coccoliths, ostracods, and conodonts, and these are described elsewhere.

Foraminifera

These are tiny unicellular organisms of the Phylum **Sarcodina** which are characterised as being amoeba-like. Members of the Order Foraminiferida produce an organic or mineralised, intra-ectoplasmic skeleton or **test** that forms chambers which are interconnected by openings or 'foraminifera'. They are first described and named in 1826 by Alycide d'Orbigny, a French palaeontologist

Figure 45 *Nummulites* a large foraminiferan from the Paleogene of Egypt.

and traveller. He spent many years exploring South America and also published a number of important monographs on the fossils of his native country.

The test is usually less that a millimeter in diameter, and its walls exhibit a variety of structures. Those that are mineralised may be laminar, being composed of three layers of calcite which may be **porous** or otherwise, while others are constructed of tiny sand grains cemented together. The test takes on a number of forms or shapes depending on the arrangements of the chambers. The simplest resemble an ear of grain and may be uniserial, biserial or even triserial. Other forms are spherical, **pyriform**, **globular** or **planispiral** in shape. On the surface the test is either smooth, highly pitted, ribbed, or may bear very fine spines.

The protoplasm or soft tissue of foraminifera is differentiated into **endoplasm** and **ectoplasm**, and resembles that of an amoeba. This is extruded out of an aperture situated on the outer chamber of the test, and can form retractile **pseudopodia** which are used in catching prey, in locomotion, and which also can secrete the test. The interior of the test is lined with an organic chitinous layer.

Foraminifera range in age from the Cambrian through to the present, and they have proved to be very useful stratigraphic zone fossils. They are also good indicators of past environmental settings, as they are known from the shallowest possible **brackish environments** typical of river estuaries to depths of up to 4000 m in the oceans. A large number of taxa became extinct at the Cretaceous/Paleogene boundary, but a number survived,

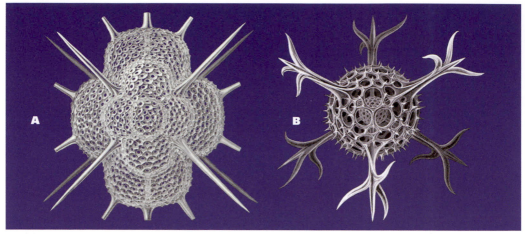

Figure 46 Radiolarians from the Atlantic Ocean: A. *Tholoma metallasson* (Haeckel, 1887); B. *Hexancistra quadricuspis* (Haeckel, 1887). The illustrations are by Ernst Haeckel (1834–1919) a zoologist and illustrator of the University of Jena in Germany, who in 1887 published a monograph of these micro-organisms collected during the H.M.S. *Challenger* oceanographic voyages.

giving rise to the modern stocks that inhabit both pelagic and benthic environments. Much of the fine-grained ooze found on the seabed in the deepest abyssal parts of the oceans is composed of radiolarians belonging to the suborder Globigerina.

Radiolarians

These are also members of the Phylum Sarcodina, and range in size from between 100 and 2000 micrometres. They live in tests of a variety of delicate and beautiful shapes composed of opaline silica. The soft tissue is divided into two regions by a central rigid pseudochitinous capsular membrane of unknown chemical composition. The exterior ectoplasm secretes the skeleton and may contain **endosymbiotic** algae which are found in those animals that live in shallow waters within the photic zone, which is the maximum uppermost 200m of water through which sunlight can penetrate,

and so algae can photosynthesise and make food; beneath that zone the water is dark. The inner endoplasm contains the nucleus and a variety of inclusions. The tissue can be drawn into outgrowths called pseudopodia that are of two types, either rigid or flexible. Radiolarians are all marine and **planktonic** in nature. They feed on small living prey such as diatoms and bacteria, which are captured using their flexible pseudopodia.

Radiolarians range from the Cambrian to the Recent and, like foraminifera, have some stratigraphical use. Given that they are composed of silica, they are most frequently found preserved in very deep water deposits at depths greater than 4000 m (Figure 46). Below this depth calcium carbonate dissolves (this is called the **calcium compensation depth**) and non-calcitic organisms obviously dominate the sediments below this zone.

2.3 Sponges

Sponges (Phylum **Porifera**) are simple multi-cellular animals that first appeared in the Precambrian. They are a loose agglomeration of cells without a well-defined skeletal support. They are mainly marine, although some freshwater sponges are known, and all are **sessile**, being attached to a substrate of some kind. Sponges prefer clean water environments, and in the past they were most diverse during times of extensive reef building when they contributed to such structures.

Sponges resemble a porous bag with an opening, or osculum, at the top (Figure 47A). The simplest body plan is that of the **ascon** sponges, and more complex were the **sycon** and **leucon** sponges, which were more efficient at filtering water (Figure 47B). The outer cell wall contains small pores called **ostia** through which water carrying nutrients can pass. Collar cells in the interior of the sponge carry flagella, which as they beat produce a pressure differential that results in water flowing through the

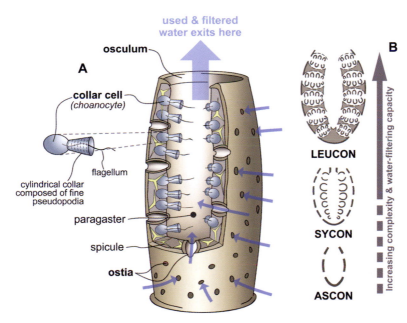

used & filtered
water exits here

osculum

A

B

collar cell
(choanocyte)

flagellum

cylindrical collar
composed of fine
pseudopodia

paragaster

spicule

ostia

LEUCON

SYCON

ASCON

Increasing complexity & water-filtering capacity

Figure 47 A. Structure of a generalised sponge and showing migration of water through the ostia into the body and out of the osculum; B. Increased complexity in sponge wall arrangement.

After Clarkson, E.N.K. 1998. *Invertebrate Palaeontology and Evolution.* Blackwell Science, Oxford.

Figure 48 A. *Vauxia gracilenta*, a Middle Cambrian bush-like sponge with cylindrical branches from the Burgess Shale of British Columbia, Canada; B. *Hallirhoa*, a large hexalobate sponge from the Greensand, Upper Cretaceous of England; C. *Siphonia*, a Cretaceous sponge that was elevated into clean seawater by means of a stem.

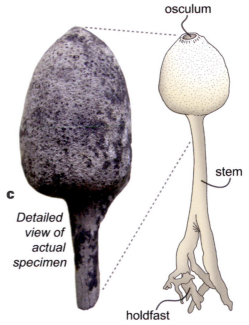

osculum

stem

Detailed view of actual specimen

holdfast

ostia, into the sponge cavity and out through the osculum. Food particles contained within the water can be extracted by the sponge. Gas exchange simply takes place by diffusion into and out of the wall cells.

Buried within the walls of the sponge is a poorly developed skeleton. In some forms this comprises organic material called spongin, while sometimes small rigid siliceous rods called **spicules** make up the skeleton. In the latter case the spongin is what produces natural sponges used in bathrooms. Spicules come in a variety of shapes such as simple pointed rods or more complex expansions resembling the small jacks used in a children's game.

The fossil record of sponges is sparse due to their poorly integrated body arrangement. In the Cambrian some early sponges such as *Protospongia* resembled a small football supported on a stem; others like *Vauxia* were bush-like (Figure 48A). Complete sponges are rare in the Mississippian, but have been reported from mudmounds in Ireland and Belgium. Isolated sponge spicules are more frequently encountered in limestones of this age. However, the finest examples of complete sponge fossils such as *Hallirhoa* (Figure 48B) are those that are preserved in the Cretaceous Greensand and Chalk. *Siphonia* was amongst the commonest taxon at the time, and its body was supported on a stem and anchored into the sediment by a **holdfast** (Figure 48C). *Coscinopora* and *Lapidospongia* formed small cup-shaped expansions. It is thought that much of the silica that formed **flints** in cavities in the Chalk was derived from the solution of sponge spicules.

Some microsponges are now only known through the borings that they made into shells and other hard surfaces.

2.4 Cnidaria

The Phylum **Cnidaria** is a diverse grouping that includes corals and sea anemones (Class **Anthozoa**), comb jellies, hydroids, and jellyfish. All but the Anthozoans have a poor fossil record. Some jellyfish fossils have been reported from the Precambrian of the Doushantuo Formation of China and the well-known Ediacaran fauna of Australia and Europe, although the biological affinities of *Charnia* remain problematic. Cnidarians are characterised by possessing **cnidoblasts**, which are stinging cells, named from the Greek *cnide* meaning 'nettle'. They are aquatic, mainly marine, although some hydroids live in freshwater environments.

Anthozoans include corals and sea anemones and consist of a soft tissue polyp which is cylindrical in shape, and made up of a **diploblastic** (two-layered) body wall. The **ectoderm**, or outer wall, carries the cnidoblasts, which when touched by a passing organism injects the prey with a venom. A network of nervous tissue is sandwiched between the ectoderm and the inner **endoderm**, which is invaginated into blade-like **mesenteries** that extend into the body cavity, or **enteron**. Surrounding the mouth are tentacles. The polyp has no anus, and no specialist circulatory or respiratory system, and so its size is controlled by the distance that gases can diffuse in and out from and to the surrounding water. Anthozoans display a **radial symmetry**.

The cnidarians are **hermaphrodite** with ovaries and testes developed within the body walls. Some spend all their lifetime as free-swimming medusae, while others are sessile polyps. Others alternate between these two states. Reproduction initially is sexual when **gametes** are formed. On fertilisation, larvae develop, which, if they settle out on a hard substrate, can grow and develop by asexual reproduction or budding. Some mature polyps then produce gametes, or else a medusa stage that can produce its own gametes, and the life cycle continues.

Anthozoans lack a medusa stage and are entirely **polyps**. The true corals, which belong to the Subclass Zoantharia, may be colonial or solitary and they have a good fossil record, as they were able to produce a calcareous skeleton. Occasionally, and exceptionally, the soft polyps are preserved. In the tabulate coral *Heliolites* from the Silurian of Canada, reddish-brown radiating lobate markings in the corallite have been recently interpreted as being the remains of polyps.

The Subclass Zoantharia is subdivided into several Orders: the **Rugosa**, **Tabulata**, and **Scleractinia**, which range from the late Precambrian to the Recent, and are characterised by the possession of a range of internal skeletal partitions (Figure 49). The coral skeleton is secreted by the ectoderm and builds up layer by layer over many generations. Painstaking examination in the 1960s of the Devonian coral *Tabulophyllum* showed that individual daily growth bands could be recognised, which demonstrated that the year at that time comprised 13 months. This was unequivocal proof that the Earth was spinning faster on its axis in the past, and that it is progressively slowing down over time.

Order Rugosa: This group is known as the tetracorals and secreted skeletons composed of calcite. They were an important and diverse group of Palaeozoic solitary or colonial corals with **septae** inserted in a four-rayed pattern. The septae are thin vertical plates which may be of varying lengths (major and minor septae)

that reflect the internal arrangment of mesenteries, and they may be joined by thin crossbars called dissepiments. Internal transverse skeletal structures are called tabulae, and these, together with the septae, provide a strong skeletal support for the corallite (Figure 49B). The soft polyp sits in a cup-shaped depression on top called the **calice** (Figure 49A). Solitary corals are broadly parallel-sided like *Siphonophyllia* (Figure 50A), cone-shaped, with circular (*Amplexizaphrentis, Cyathophyllum*), square

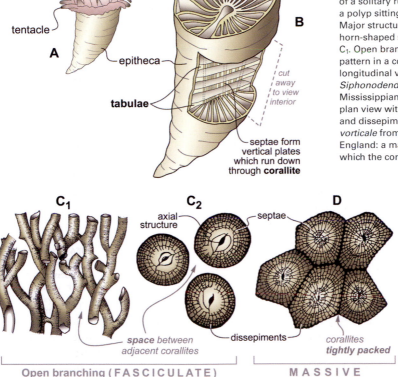

Figure 49 A–B. Reconstruction of a solitary rugose coral with a polyp sitting in the calice; B. Major structural elements of a horn-shaped solitary rugose coral; C$_1$. Open branching (fasciculate) pattern in a colonial rugose coral in longitudinal view; C$_2$. Corallites of *Siphonodendron martini* from the Mississippian of England seen in plan view with arrangement of septa and dissepiments; D. *Lithostrotion vorticale* from the Mississippian of England: a massive rugose coral in which the corallite walls are touching.

C$_1$ after Clarkson, E.N.K. 1998. *Invertebrate Palaeontology and Evolution.* Blackwell Science, Oxford. C$_2$ and D after Nudds, J.R. 1980. An illustrated key to British Lithostrotionid corals. *Acta Palaeontologica Polonica* **25**, 385–394.

or semi-circular (*Calceola*) cross-sections. Colonial (compound) corals consist of many adjacent polyps which make up the **corallite**. These may be touching, and so form massive colonies (*Lithostrotion, Acervularia* (Figures 49B, 50B)), or fasciculate in pattern where gaps occur in between adjacent polyps as in *Siphonodendron* (Figure 49C$_{1-2}$). The first record of the Rugose corals is in the middle Ordovician, after which they diversified rapidly, reaching

Figure 50 A. *Siphonophyllia gigantea*, a solitary rugose coral from the Mississippian of Co. Sligo, Ireland showing (left) a cross-section with radial septae, and (right) a longitudinal section with horizontal tabulae; B. *Acervularia*, a colonial rugose coral from the Silurian of England; C. *Halysites*, the colonial tabulate 'chain coral' from the Silurian of Gotland, Sweden; D. Scleractinian coral from the Matmor Formation, Jurassic of Makhtesh Gadol, Israel.

150 genera by the middle Devonian. An extinction saw diversity plummet, although through the early Mississippian diversity increased again. In the Pennsylvanian and Permian they slowly declined and the group became extinct at the end of the Permian.

Order Tabulata: Exclusively colonial corals with a small corallite composed of calcite. They have prominent **tabulae**, and septae are absent or very short, equisized and usually 12 in number. In some taxa the septae occur as short septal spines only. Colonial integration between the polyps is enhanced by mural pores that may allow the transfer of nerve impulses, or gases from polyp to polyp. On account of there being few skeletal features, corallites are usually simple. Easily recognisable taxa include *Halysites* (Figure 50C), the 'chain coral', *Michelinia* and *Favosites*. The Tabulate corals appeared during the Lower Ordovician and diversified until the middle Devonian, when an extinction event wiped out many genera. They never fully recovered and became extinct at the end of the Permian.

Order Scleractinia: These include all the modern corals, and appeared at the end of the Lower Triassic. Their skeleton is composed of aragonite, and form either solitary or colonial corals (Figure 50D), usually with radial symmetry. Septae are inserted in cycles of six, and so the group is widely known as 'hexacorals'. Solitary corals are usually cone-shaped or straight. Colonial or compound corals are produced by repeated asexual budding; intercorallite tissue may be developed between adjacent polyps. The origins of the Scleractinian corals and their relationship with earlier Orders is unclear due to gaps or non-preservation of antecedents in the fossil record. One remarkable genus, *Kilbuchophyllia*, from the Middle Ordovician of Scotland, has scleractinian-like septal insertion and is classified in its own Order, but there is a gap of 230 million years between it and the Scleractinians, and so any possible linkages are clouded by time. It is clear, however, that the Scleractinians rapidly diversified to fill the environmental niches vacated by the earlier Rugose and Tabulate corals. Today over 200 genera are found in modern oceans, but they are quite temperature and salinity dependent, and changes in modern climatic conditions continue to be a worry to ecologists.

Scleractinian corals were able to cement themselves firmly to the **substrate**, and some became major **reef-builders**. These are the **hermatypic** corals that live in a **symbiotic** relationship with **zooxanthellae**, or photosynthetic algae. As such, they are restricted to the shallow photic zone, and favour warm (18–29 degrees centigrade), clear waters between 30° North and South of the Equator. The algae provided food which allowed the coral to grow three times faster than it would without this symbiotic relationship. Given their ability to grow fast and to produce a great deal of aragonitic skeleton, the corals throughout the Mesozoic and Cenozoic provided a large volume of carbonate sediment now preserved in young limestones. Coral reefs are major sources of nutrients and are characterised by a wide diversity of attendant species such as fish. Modern parrotfish are responsible for producing copious volumes of carbonate sediment produced as a result of their grazing the coral skeletons, which they digest and defecate. Today cool water corals (**ahermatypic**) such as *Lophelia* live at depths down to 6000 m in the Atlantic Ocean and elsewhere. These do not have associated photosynthetic algae.

2.5　Bryozoans

Bryozoans (Phylum **Bryozoa**) are marine or freshwater **colonial** invertebrates that first appeared in the Ordovician, and which extend to the present day. Like some sponges and corals, they form colonies of various shapes and sizes, usually with calcite skeletons, and so have a high fossilisation potential. Approximately 20,000 fossil species and 5000 living species are now known. They are common as reef-dwellers, both today and in the past, especially in the Ordovician and the Carboniferous. Marine bryozoans usually inhabit shallow environments from the poles to the Equator, usually at depths of less than 50 m, although some deep-water forms are known. Palaeozoic forms are well known from the Ordovician of near Cincinnati, Ohio; the Devonian of Germany and the Czech Republic; the Mississippian of Illinois, the UK and Ireland; the Permian of north-east England, Italy, and Australia. Mesozoic bryozoans are prominent in the Cretaceous of England and Denmark, while Paleogene faunas include those of the Pliocene Coralline Crag in Suffolk, which contains 150–200 species. These resemble the bryozoan-rich modern carbonates currently forming off the coast of southern Australia.

The individual bryozoan animals typically measure 1 mm or less in diameter and are called **zooids**. Most are adapted for feeding and are called autozooids (Figure 51), but some, called heterozooids, have other functions in the colony, such as for defence or cleaning. The autozooid comprises a **polypide** (the feeding unit) and epithelial tissue which secretes the skeleton of a tubular or box-like chamber called the zooecium. The polypide consists of a **lophophore** (tentacle crown), a U-shaped gut and reproductive organs. Lophophores composed of up to 30 tentacles are protruded (everted) through apertures for feeding, or retracted by means of a retractor muscle while at rest (Figure 51). In some taxa apertures are protected by opercula or lids.

The defence heterozooids include avicularia that use a sharp mandible to snap and bite predators, which include pycnogonids (sea spiders) and nudibranchs (sea slugs). Others form bristles called vibracularia which, when touched, send a nerve impulse to the animal, which then retracts if feeding. The signal is then sent to neighbouring zooids, which react in the same way.

Beating of cilia on the tentacles in autozooids creates **inhalent** water currents which carry food particles to the mouth, and exhalent currents which carry waste away. In **reticulate** meshwork colonies often seen in the Order Fenestrata (*Fenestella*) the waste-water passes through openings in the meshwork called **fenestrules** (Figure 52A). In larger massive colonies such as those frequently seen in the Order Trepostomata (*Hallopora*) where there is no meshwork, the skeleton may be developed into regularly placed **monticles** or hummocks which may be star-shaped as in the Ordovician genus *Constellaria* (Figure 52B) or elongate in *Spatiopora* (Figure 54C).

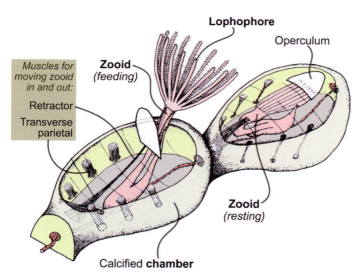

Lophophore

Operculum

Zooid (feeding)

Muscles for moving zooid in and out:

Retractor

Transverse parietal

Zooid (resting)

Calcified **chamber**

Figure 51 Generalised morphology of a Recent gymnolaemate bryozoan autozooid with the lophophore everted in the feeding position (left) and retracted into the zooecium (right).

After McKinney, F.K. 1991. *Exercises in invertebrate paleontology.* Blackwell Scientific Publications, Oxford.

Figure 52 A. *Fenestella*, a fenestrate bryozoan with large fenestrules forming a coarse meshwork through which water could flow. Mississippian of Hook Head, Co. Wexford, Ireland; B. *Constellaria*, a trepostome bryozoan with star-shaped monticles, from the Kobe Formation, Ordovician of Cincinnati, Ohio, USA; ⊃

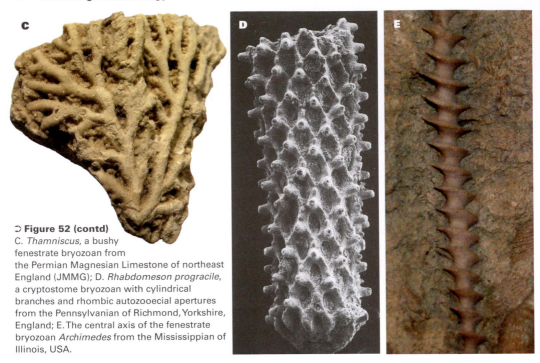

⊃ **Figure 52 (contd)**
C. *Thamniscus*, a bushy
fenestrate bryozoan from
the Permian Magnesian Limestone of northeast
England (JMMG); D. *Rhabdomeson progracile*,
a cryptostome bryozoan with cylindrical
branches and rhombic autozooecial apertures
from the Pennsylvanian of Richmond, Yorkshire,
England; E. The central axis of the fenestrate
bryozoan *Archimedes* from the Mississippian of
Illinois, USA.

Close examination of these show that they lack autozooecia, which instead are developed in the hollows between. The monticles act as 'chimneys' for the escaping exhalent current.

Bryozoans are hermaphrodite, and their sex organs are housed in the zooecial chamber. In order to reproduce these produce gametes sexually, which fuse to form a larva. This may be either released immediately into the sea or it may be transferred to an inflated **brood chamber** called an ovicell that is often situated adjacent to the aperture. There the larva can mature for a while. Larvae are of two types: **lectotrophic** and **planctotrophic**. The former settle close to the adult colony, whereas the latter bear cilia and a food source which allows them to live longer and swim further from the parent. **Cosmopolitan species** generally have planctotrophic larvae. Once the larva settles on a suitable substratum it develops into the first zooid, called ancestrula, and from this all the succeeding zooids are budded asexually.

Except in freshwater bryozoans (Class **Phylactolaemata**) where there is no calcification, individual animals live in calcified chambers that make up the zoarium of the colony. Basically there are two forms: (A) in the Class **Stenolaemata** pattern where the living chambers are generally simple and cylindrical in shape and (B) in the Class **Gymnolaemata** pattern where chambers are box-like. A collection of chambers combine to make up a colony, which comes in many shapes and sizes. Typically colonies reach 100 mm in diameter, but occasionally some freshwater colonies may be some metres in diameter. The Stenolaemata contained five orders, four of which became extinct or severely decimated at the end of the Permian; the modern bryozoans evolved from the Cyclostomata during the Mesozoic.

Most bryozoans are sessile, permanently cemented to a firm substrate such as seaweed, shells of various groups or corals. One modern form is an exception to this in that it can move across the sediment surface by coordinated action of its vibracula which act like walking legs. Some have a boring habit. Encrusting colonies are simple, with a runner-like morphology, or are composed of single or multi-layered sheets of zooids. Commonly colonies are erect, which allows them to feed from cleaner and faster flowing water above that close to the sediment of the sea floor. They may be simple expansions of dividing branches (Figure 52C) or complex and tree-like with many branches. Branches may be cylindrical (Figure 52D), either thick (ramose) or thin (dendroid), jointed or unjointed. They may be flattened with zooids budded either side of a median lamina (bilaminate) or with zooids on one surface only (unilaminate). Many unilaminate bryozoans form a reticulate pattern of branches joined by crossbars called dissepiments. Some bryozoans form dense small domed colonies, while others take on the shape of the shell or other substrate encrusted. Usually erect colonies become broken into smaller branch fragments after death, and are rarely found preserved intact in the fossil record. Throughout the geological past many particular morphologies, including reticulate or **monticulate** colonies, have evolved repeatedly, although the morphology of the individual chambers that comprise these colonies is radically different in the Palaeozoic than in today's forms.

Classification

Class Phylactolaemata: Freshwater bryozoans with no calcification. Fossil record poor: some resting stages (statoblasts) known from Triassic sediments in South Africa.
Class Stenolaemata: Marine bryozoans with tubular zooids with calcified walls.
Order Cystoporata [Lower Ordovician–Triassic]: Erect or encrusting colonies, of long, simple zooecial chambers with basal diaphragms. Vesicular tissue and/or hood-like lunaria on apertures diagnostic.
Order Cryptostomata [Lower Ordovician–Permian]: Cylindrical or bifoliate erect colonies (Figure 52D). Autozooecia budded from axis or median lamina.
Order Fenestrata [Lower Ordovician–Permian]: Colonies erect, formed of thin dividing branches (Figure 52C) or reticulate meshwork (Figure 52A) arranged in fans or cones (*Fenestella s.l.*) or coiled around an axis (*Archimedes* (Figure 52E)). Autozooecia open onto one face only.
Order Trepostomata [Lower Ordovician–Upper Triassic]: Colonies encrusting or erect; branches up to 5 mm thick (Figure 52B). Zooecial chambers long, many diaphragms; walls thin in centre (endozone) thicken towards surface (exozone).
Order Cyclostomata [Lower Ordovician–Recent]: Colonies encrusting or erect. Autozooecial chambers long. Communication pores between zooids and brooding chambers in most genera.

Class Gymnolaemata: Mostly marine bryozoans with cylindrical or flattened zooids.
Order Ctenostomata [Lower Ordovician–Recent]: Skeleton gelatinous or membranous. Some species have a boring habit.
Order Cheilostomata [Upper Jurassic–Recent]: Colonies erect or encrusting, composed of box-like zooecia. Avicularia common.
Suborder Anasca: Frontal wall membranous or chitinous, uncalcified.
Suborder Ascophora: Calcified frontal wall, perforate or imperforate.

2.6 Molluscs

The molluscs (Phylum **Mollusca**), from the Latin *molluscus* meaning 'soft', are a group of superficially morphologically varied organisms including cephalopods (squids, nautiloids, and octopus), bivalves (oysters, cockles, mussels and clams), gastropods (snails and slugs) and less well-known groups such as chitons. Many of these today end up on the tables of restaurants and are important sources of food for many people. All molluscs possess common features such as a muscular foot, a **mantle** that drapes over the shell, which it secretes either in calcite or aragonite, and a visceral mass that contains the vital organs such as the heart, stomach, and gonads. The mantle is also invaginated to produce the **mantle cavity**, a water-filled area that holds the gills that are used for respiration, and into which the intestine empties via the anus. With the exception of the gastropods, which are hermaphrodite, most molluscs have separate sexes.

Over 70,000 modern species are found in wide-ranging habitats from terrestrial to marine settings, where they have adopted a variety of feeding strategies. Some are swimmers, others benthic grazers, some are attached and sessile suspension feeders, while a large proportion of bivalves are **infaunal** deposit and suspension feeders that live at different depths within the sediment. The Colossal Squid (*Mesonychoteuthis hamiltoni*), which lives in cold southern hemisphere waters, measures up to 14 m long and is thought to be the largest **invertebrate** living today. Mollusc shells are often prized by collectors on account of their enormous beauty, colour, lustre and range of morphologies. This said, collectors need to remember that removal of living animals, and indeed, of certain shells, may be in contravention of both local and international laws. The study of molluscs is important enough to merit its own term, '**malacology**', and the study of their shells is called '**conchology**'.

Most molluscs are now considered to have originated at various times in the Cambrian from a common ancestor, the archimollusc or 'ancient mollusc'. The notion of such an animal was first raised by Ray Lankester, an English zoologist, in 1883, and for some time the concept remained theoretical. In the 1920s and more recently, fossils of early members of the phylum have been recovered: *Latouchella* from England and *Yochelcionella* from Canada, Greenland, China and Pennsylvania, and elsewhere. These are grouped together into the Class Helcionelloida and appear to be more closely related to the gastropods than to the other mollusc classes. *Neopilina*, which was until recently thought to be the only modern example of a similar primitive mollusc, was discovered in 1953 inhabiting very deep water; but recent re-examination and comparison with fossil forms shows it to differ from them significantly. These early molluscs that were derived from the archimollusc broadly resembled a limpet, albeit only a few millimetres in diameter, having a cap shell which was rather flat or cone-like, and recumbent either **anteriorly** or **posteriorly**. Some groups such as the Scaphopods appeared later during the Ordovician.

It has been estimated that at least 40,000 fossil molluscs have been described.

2.6.1 Cephalopods

Cephalopods (Class **Cephalopoda**) are the most advanced of all molluscs and include modern-day squids, octopus, cuttlefish and the pearly nautilius (*Nautilus*), and the fossil nautiloids and ammonoids. They have internal chambered shells (phragmocone)

that are gas-filled, and all the chambers linked by a tube called the **siphuncle** which carries a thread of flesh from the animal that lives in the last enlarged chamber called the **body chamber** (Figure 53). This passes through the septae via the septal neck, which is the part of the septa that is bent backwards. The gas-filled chambers make the shells buoyant. In many taxa the first chamber or **protoconch** can be seen at the apex or centre of shells; the protoconch hatched out of an egg. The walls that divide the chambers are called septae and may be either simple curved walls, or rather elaborate. Cephalopods have a properly defined head with elaborate sense organs

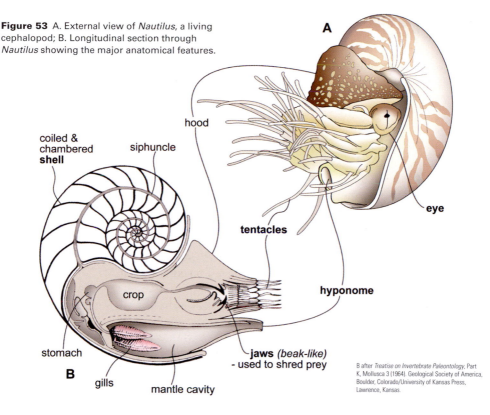

Figure 53 A. External view of *Nautilus*, a living cephalopod; B. Longitudinal section through *Nautilus* showing the major anatomical features.

A

coiled & chambered **shell**

siphuncle

hood

eye

tentacles

crop

hyponome

stomach

B

gills

jaws *(beak-like)* - used to shred prey

mantle cavity

B after *Treatise on Invertebrate Paleontology*, Part K, Mollusca 3 (1964). Geological Society of America, Boulder, Colorado/University of Kansas Press, Lawrence, Kansas.

including eyes, a mouth, tentacles, and move by propelling water at high speed out from the mantle cavity via the **hyponome**, which is a tube situated near the lower part of the ventral side of the animal (Figure 53). **Gills** for respiration are situated in the mantle cavity. Today cephalopods generally live in open oceanic settings and are active at night, when they are found in surface waters. By day they alter their buoyancy and sink into deeper waters. Studies of the strength of cephalopod shells under increasing pressure have shown that they could live at depths of approximately 200 m, but most modern forms are more frequently found within the top 50 m.

Cephalopods were equipped with jaws that comprised a pair of **aptychi** which formed a hood over the mouth, and a radula situated just inside it. They most likely fed on fish. Early cephalopods would have been amongst the major predators in the oceans, but with the rise of the fishes in the Devonian and the diversification of marine reptiles in the Mesozoic, they became as much prey as predator. There are some good examples in the fossil record of ammonites with fatal puncture marks on the shell, which match nicely to the dental arrangement of **mosasaurs**. Occasionally fossils are found with traces of colour remaining on the shell – it is probable that cephalopods were highly coloured with a variety of stripes, which would have helped camouflage the animal as it swam.

The Class Cephalopoda includes six subclasses, the three of which most commonly found as fossils being treated here.

Nautiloids (*Subclass Nautiloidea*) first appeared in the Cambrian and today one genus *Nautilius* is extant (Figure 53). The shells were generally straight (orthoconic) (Figure 54A) or slightly curved (cyrtoconic) in the older orders found in Cambrian to Silurian strata, but become loosely to tightly coiled in later representatives more commonly found in the late Palaeozoic. Only coiled nautiloids are present today. The outer shell may be smooth, striated or heavily ribbed. The internal septae are generally curved concavely towards the posterior of the shell, or apex if straight shelled, and where these septal walls meet the surface shell they produce a simple **suture**. In many fossils this suture can only be seen if the outer shell is not preserved, or chipped away by a fossil preparator. The shape of the suture is a useful taxonomic feature used in classification. Within many genera additional skeletal material called **cameral deposits** are found on the bottom of chambers. This helps the animal to keep correctly orientated in the sea water. Without such deposits it would be difficult for the animal to stay upright. Nautiloids ranged in size from a few centimetres to a metre in length. On death the chambers filled with water and the shells would have sunk to the seabed. Many orthoconic shells have been found aligned, which suggests that they have been moved by water currents following death. In rare examples trilobites have been found inside the body chamber – they probably crawled in when it was empty and became smothered by overlying sediment (Figure 54B). Nautiloid shells were frequently encrusted by epizoans such as brachiopods and bryozoans. Sometimes, where only one side of the shell is encrusted, this proves that encrustation occurred after death, but in some instances the epizoan occurs over all the shell, which demonstrates that the encrustation took place while the nautiloid was alive. The bryozoan

Figure 54 A. *Rayonnoceras espeyensis*, an orthoconic nautiloid seen in longitudinal section showing a large siphuncle. Mississippian of Castle Espie, Co. Down, Ireland; B. Two *Alcymene pisellaris* trilobites that sheltered in the body chamber of the nautiloid *Polygrammoceras bullatum* from the Silurian of Wales; C. *Spatiopora*, a bryozoan with elevated elongate monticles encrusting an orthoconic nautiloid from the Waynesville Formation, Ordovician of the Cincinnati area, Ohio, USA.

Spatiopora from the Ordovician of Cincinnati, Ohio actually arranged its feeding zooids in rows to maximize feeding while the nautiloid was swimming (Figure 54C).

Ammonites and Goniatites (*Subclass Ammonoidea*) can be distinguished from nautiloids, from which they evolved in the early Devonian, by possession of a more complex suture. They also differ in that they have a tightly coiled plane-spiral phragmocone that may be evolute (part of all of the whorls can be seen) or involute (the outer whorl covers all the earlier inner whorls), and a marginal siphuncle.

Goniatites (Figure 55A–B) are characterised by the possession of a suture that consists of a series of curved **saddles** and pointed **lobes** (Figure 55B), as well as a generally globose shell. The siphuncle is median and the shell walls are moderately thick. They reached their maximum diversity during the Upper Palaeozoic and in the Pennsylvanian in particular, when at least nineteen genera are known. As they evolved rapidly, and as each species had a short duration, they have proved to be very useful marker or index fossils. Many Pennsylvanian successions consist of terrestrial sediments that were deposited close to the shoreline, and interspersed throughout these

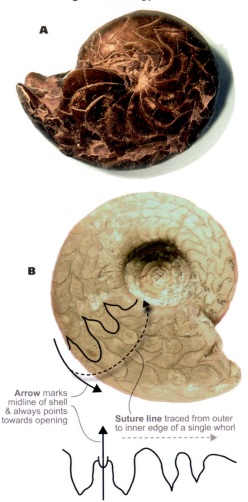

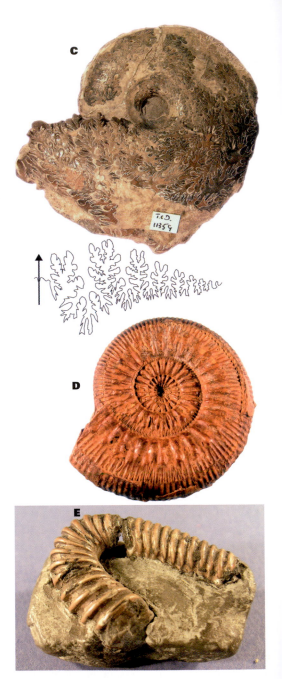

Arrow marks midline of shell & always points towards opening

Suture line traced from outer to inner edge of a single whorl

Figure 55 A. *Goniatites*, a pyritised goniatite with simple sutures from the Pennsylvanian of Rush, Co. Dublin, Ireland; B. *Uraloceras*, a goniatite from the Permian of Aktiubinsk, Kazakhstan, with a tracing of its suture; C. *Phylloceras heterophyllum*, a Lower Jurassic ammonite from the Mickleton Railway Tunnel, Co. Durham, England. The outer shell has been removed to reveal the complex sutures (tracing); D. *Stephanoceras humphriesianum*, a nodose ammonite from the Inferior Oolite, Middle Jurassic of England; E. *Hamites*, a Cretaceous ammonite heteromorph.

B after Ubukata, T., Tanabe, K., Shigeta, Y., Maeda, H. and Mapes, R.H. 2010. Eigenshape analysis of ammonoid sutures. *Lethaia* **43**, 266–277.

thick sequences are thin marine bands that represent incursions of the Rhaetic Ocean due to sea level fluctuations. The marine bands can be identified as marine due to the presence of goniatites, and many of them can be correlated over huge distances, as they contain distinctive goniatite species.

The ammonites are distinguished from goniatites in their development of a very complex suture (Figure 55C), thinner walls than in nautiloids, and the siphuncle is positioned close to the outer portion of the shell. The shell is triple-layered with a middle mother-of-pearl layer covered by thinner aragonitic layers. Preservation in some instances of this central nacreous layer has produced some beautiful specimens such as those in the Lower Jurassic Marston Marble of Marston Magna, Somerset in England (Figure 1). The surfaces of shells are frequently highly ornamented, with ribs or rows of small to large nodes and knobs as in *Stephanoceras* (Figure 55D). Studies of models in flume tanks subjected to water currents showed that the more ornamented shells were more stable in the water than their smoother counterparts. Some ammonites developed small lateral projections called **lappets**, which may have protected the tentacles. Ammonites also exhibited **sexual dimorphism**. The females were larger than the males, but otherwise the sexes were morphologically similar. The largest recorded ammonite is *Parapuzosia* from the Cretaceous, which reached a staggering diameter of 2.5 m.

Ammonites first appeared in the lower Triassic, but all of the post-Triassic forms were derived from just one genus, *Phylloceras* (Figure 55C). Rapid diversification occurred in the Jurassic and Cretaceous, which has resulted in their being most useful zone fossils and important for global correlation (*see* Section 1.6.4); but ammonites fell victim to the end-Cretaceous extinction event and are not found in post-Mesozoic successions.

Some highly unusual shell shapes developed in the ammonites at several points in their history, and these odd forms are collectively called **heteromorphs** (Figure 55E). Some externally resembled gastropods, whereas in *Macroscaphites* the shell became partially unrolled with the body chamber turned backwards towards the rest of the shell.

Belemnites are a member of the Subclass Coleoidea that include the modern-day cuttlefishes, octopuses and squids. Belemnites, which are extinct, lived between the Triassic and Cretaceous, and are characterised by an internal chambered shell called the **phragmocone**, situated at the rear of the animal, which is surrounded by a bullet-shaped guard that can reach 20 cm in length. In most cases this is the only part preserved, but in some rare specimens such as in *Cylindroteuthis oweni* from the Oxford Clay, Upper Jurassic of Wiltshire, England (Figure 56) even the dark ink sac has survived. The anterior portion of the animal comprised a soft body containing all the vital organs, a head with large eyes, and tentacles at the very front, used to aid feeding. Guards are often encrusted by **epibionts** – most frequently these are bryozoans – which shows that they were encrusted after the animal died. Belemnites can be found concentrated into what are termed 'Belemnite Battlefields', and this tells palaeontologists that the guards were transported after death by water currents. These readily take up a preferred orientation that reflects the water movement direction. In

Figure 56 *Cylindroteuthis oweni*, a fine specimen showing the bullet-shaped guard, the triangular phragmocone, and the dark ink sac from the Oxford Clay, Upper Jurassic of Christian Malford, Wiltshire, England.

some extreme cases concentration has taken place in the stomachs of marine sharks; one such animal discovered in Germany had eaten over 250 animals before it died. Due to their diversity and often short ranges, belemnites have been used as index or zone fossils for the latter two periods of the Mesozoic.

2.6.2 Gastropods

Gastropods (Class **Gastropoda**) are one of the main divisions of the Phylum Mollusca. They include snails, which bear a coiled or uncoiled calcareous shell (Figure 57), and others, such as slugs, which have no hard parts. Typical of the Class Gastropoda are the land and freshwater snails, the many conchs, whelks, limpets and periwinkles. They first appeared in the fossil record in the Lower Cambrian, and throughout the Palaeozoic they remained entirely marine. However, in Mesozoic and Cenozoic times large numbers of them became adapted for life in fresh waters and on the land. A large majority of the group remained in the sea. The average size of the shells of the group is approximately 25 mm in length or diameter,

but fully grown adults of different kinds range from 0.5 mm to approximately 60 cm.

The modern-day land snails, *Helix*, and the shallow-water marine *Buccinum* (Figure 55) are familiar examples of gastropods. In each case the animal is elongate with a flattened 'foot' on which it moves, and it carries a spirally coiled shell on its back. The tip of the spiral shell is the earliest-formed part of the shell, called the protoconch, and it points backwards, and the opening into the largest, last-formed turn of the shell is in a forward position, directed downward. The last whorl of the shell comprises the **body chamber** while the earlier whorls are referred to as the spire. The animal can withdraw into the shell and many species protect themselves by means of an **operculum** which closes off the aperture as it retreats into the shell. The coiling of shells may be sinistral (left-handed) where the aperture appears on the left-hand side, or dextral where it is on the right side when viewed. In many species, including *Athleta* from the Eocene, the aperture has a canal-like lip which carries the inhalent siphon that is used as a conduit for clean water.

Figure 57 A. The living gastropod *Buccinum undatum*, the Common Whelk, showing protrusible soft parts; B. An empty shell of *Buccinum* with shell terminology outlined.

A after *Treatise on Invertebrate Paleontology*, Part I, Mollusca 1 (1960). Geological Society of America, Boulder, Colorado/University of Kansas Press, Lawrence, Kansas.

B after Clarkson, E.N.K. 1998. *Invertebrate Palaeontology and Evolution*. Blackwell Science, Oxford.

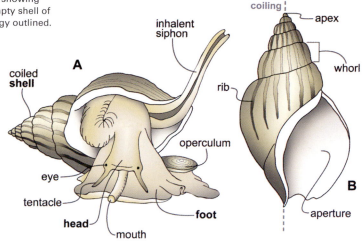

The head is the mobile, anterior part of the body that bears the mouth, eyes and one or two pairs of tentacles. Situated just inside the mouth is a long muscular movable rasping mechanism that is composed of many minute horny teeth arranged in transverse rows on a tough, flexible ribbon. This mechanism, called the **radula**, functions for grazing food from rocky surfaces (Figure 101D) and tearing food into particles, and has been used by taxonomists as an aid to classifying living gastropods. Eyes are borne at the tips of the tentacles of some gastropods, but in others they are situated near the base of the tentacles. The gastropods are hermaphrodite and respire by means of **gills** situated within the body.

The calcareous or aragonitic shell is the structure on which palaeontological study of the group must be based. Most gastropod shells show a wide range of form and structure which can be used in identifying species and genera. Shells of gastropods may be divided into two groups in general: those with little or no coiling (**patelliform**) such as the modern-day limpet *Patella*, and those which are partly or very distinctively coiled. Of the coiled forms, some show a flattened test (**planispiral**) as in *Straparollus* or *Phanerotinus* (Figure 58A) from the Mississippian, while in others the coiling can be drawn out into a high spire as in *Turritella* (Figure 58B). Others, such as the modern top shell, have a **trochiform** shell with a pyramidal spire and a flat base.

The majority of gastropods are aquatic, and most species live in shallow marine waters. Some have become adapted to brackish and freshwater environments, and two groups, by modifying their breathing apparatus, have invaded the land. Today gastropods are found in a variety of environments. *Phanerotinus* developed large triangular flattened flanges along its lateral margins, and these suggest that the shell lay on a muddy substrate with the flanges distributing the weight so that the animal did not sink into the mud.

Figure 58 A. *Phanerotinus*, a giant flanged Mississippian gastropod known only from England and Ireland; B. *Turritella carinata* gastropods from the Pliocene of Cyprus.

2.6.3 Bivalves

Bivalves (Class **Bivalvia**), or Lamellibranchs or Pelecypods, as they are also known, include cockles, mussels, and oysters and consist of a shell of two-hinged **valves** (Figure 59A$_1$). Superficially they resemble brachiopods in this respect, but are morphologically and symmetrically quite different. In most genera the valves are mirror images of each other when viewed from their **anterior** or **posterior** sides, but in some groups, such as the oysters and the **rudists**, this is not the case, and one valve is usually larger than the other. The smallest bivalves were a few millimetres in diameter and the largest, some Permian giant clams, reached over 1 m in width.

On the exterior surface growth lines are often seen developed in concentric circles away from the beak or **umbo** situated on the dorsal margin (Figures 59A$_1$, 60A). The two valves are hinged along their dorsal margin with a series of **ligaments** and a **tooth-and-socket** arrangement, the latter being a useful aid to the classification of the group. The two

valves are also joined together by a pair of adductor muscles whose point of attachment is regularly preserved as a droplet-shaped scar. When these muscles relax the ligament at the hinge line expands and the valves open along their ventral margin, thus allowing water and food to be drawn in.

The soft tissues of the bivalves include a gill structure used for respiration, which is contained within the **visceral mass**, a mouth leading to an alimentary canal that ends in the anus, and a fleshy foot (Figure 59A$_3$). On the internal side of the valves the **pallial line** or sinus runs between the adductor muscle scars, and this marks the point of attachment of the mantle which is responsible for the precipitation of the shell. In some taxa the pallial line is inflected on its posterior side, where it accommodates **inhalent** and **exhalent** **siphons** (Figure 59A$_2$).

Bivalves have a long geological history, first appearing in the Lower Cambrian, but they remained rather insignificant members of the Palaeozoic biota. Following the mass

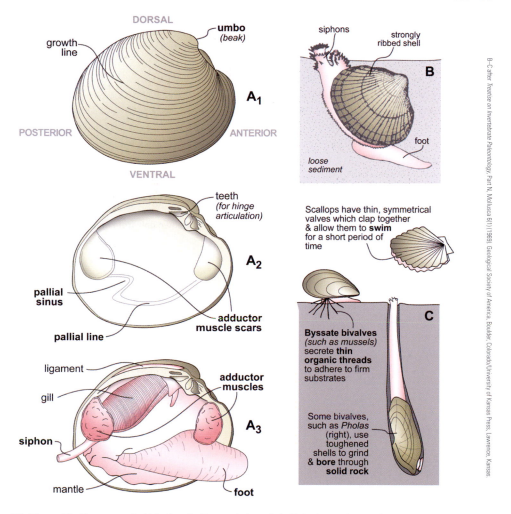

B–C after *Treatise on Invertebrate Paleontology*, Part N, Mollusca 6(1) (1969). Geological Society of America, Boulder, Colorado/University of Kansas Press, Lawrence, Kansas.

Figure 59 The cockle *Venus*, a typical bivalve. A₁. External view of shell; A₂. Internal view; A₃. Major anatomical features; B. The shallow burrowing bivalve *Cardium edule* with siphons extending to the sediment surface; C. *Pholas dactylus* in its boring with the shell attached by a sucker-like foot, above which attached to the sediment is a byssate bivalve, and swimming in the water a pectinid bivalve.

extinction at the end of the Permian, the brachiopod stocks that favoured environmental niches in shallow-water continental shelf settings were devastated. The bivalves took advantage of this and diversified from the early Mesozoic, displacing the brachiopods from their former niches. They were conspicuous elements of Mesozoic and Cenozoic biotas and have been found in a range of sedimentary rocks.

Examination of the shape and morphology of fossil bivalves, together with a comparison of modern shells and their life habits, has illustrated the versatility of this group. Many are infaunal burrowers (Figure 59B), some borers; others may be **epifaunal** living on the seabed, while others are active swimmers (Figure 59C). The infaunal burrowing bivalves exploited this niche to escape from predators, but in order to feed they evolved siphons or elongated tubes that reached the sediment surface and through which water laden with food was taken in through the inhalent siphon and expelled via the exhalent siphon. Boring bivalves have adapted to high energy conditions by protecting their shells by boring into hard substrates such as rocks or timber. The anterior side of the valves of the **endolithic** *Pholas*, which bores into limestone, are covered with hard nodes that are used to excavate the boring. These borings generally are flash-shaped, which means that once full-sized the adult cannot escape (Figures 59C, 102B). The epifaunal bivalves include *Pinna* (Mississippian to Recent) and the modern mussel *Mytilus* which is attached to the substrate by a series of threads called byssus (Figure 59C). Many oysters are physically cemented to the substrate, but other members of the oyster family including *Gryphaea* simply lie in the mud with the weight of their heavy convex lower valve simply holding them in place (Figure 60B). Scallops are free-swimming pectinid bivalves characterised by having two very flat valves with a long hinge-line extended into 'ears' on either side of the beak (Figures 59C, 60C). They 'swim' by clapping the valves together, which ejects the water out of the 'ears' and propels them in a wavy trajectory through the water. Some pectinids are not swimmers but

are epifaunal, and these have an **umbonal angle** of less than 105 degrees, while the swimmers have an umbonal angle of greater than 105 degrees.

The most unusual bivalves are the rudists (Figure 60D) which evolved during the Mesozoic and are characterised by having two different valve sizes. *Hippuritella*, which can be regarded as being a typical rudist of an atypical group, was conical in shape with its right valve forming the cone and its left valve forming a cap-like apparatus on top. Unlike their more conventional counterparts, rudists were important reef-dwellers, particularly in the mid-Cretaceous, but unlike many, they disappeared at the extinction event at the end of that geological period.

Bivalves are notoriously difficult to classify. Most taxonomists used a combination of the hinge dentition, the shape of the adductor muscle scars, the shape of the pallial line, and the shell mineralogy to distinguish groups at various different taxonomic levels. Shell shape is also a useful guide, but reliance on it alone can cause difficulties, as similar shaped shells have evolved in distinct families of bivalves at different points in the fossil record.

2.6.4 Other molluscs

Chitons (*Class Polyplacophora*) are a class of marine molluscs that include nearly 1000 extant species and which has a reasonable fossil record that extends back to the Upper Cambrian. These animals possess an elongate oval-shaped body usually less than 10 cm long, covered with eight plates which suggested their common name, 'coat-of-mail' shells. These plates cover the unsegmented body, which has a flattened lower foot on which the

Figure 60 A. *Plagiostoma giganteum*, a large bivalve from the Jurassic of England; B. *Gryphaea* (*Bilobissa*) *dilatata*, an epibenthic bivalve with a wide shell which sat on the muddy seabed. Oxford Clay, Upper Jurassic of England; C. *Chesapectin jeffersonius* scallops from the Pliocene deposits around the Chesapeake River, Yorktown, Virginia, USA. These 'Virginia Clams' were named for Thomas Jefferson and were designated the State Fossil for Virginia in 1993. One valve is encrusted with some sediment and some *Balanus* barnacles; D. Rudist bivalves from the Cretaceous of the Omani Mountains, United Arab Emirates.

animal moves. They live in a variety of water depths as far down as 7000 m, but are most frequently found in high energy nearshore areas where they cling to rocks and graze the surfaces using their radula. On death the plates tend to disarticulate, and those examples found articulated represent adults that had been rapidly buried. Several Palaeozoic genera are known from Scotland, Ireland, Bohemia, Illinois and Texas.

Scaphopods (*Class Scaphopoda*) are the tuskshells that have a long cylindrical tapering shell with openings at either end (as in the

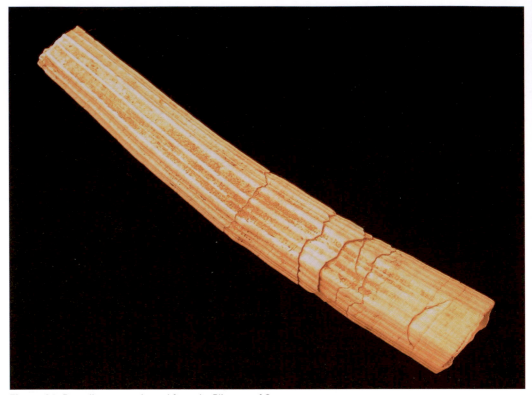

Figure 61 *Dentalium*, a scaphopod from the Pliocene of Cyprus.

Palaeozoic genus *Dentalium* (Figure 61) which has longitudinal ridges on its shell). In life position the anterior portion of the shell is embedded in the sediment while the apex sticks up into the water. They ranged from the Ordovician to Recent.

Rostroconchs (*Class Rostroconchia*) are oddly shaped molluscs that are often mistaken for bivalves but in fact are structurally distinct. The valves have a lateral expansion that resembles a long beak. They first appeared in the Cambrian, became conspicious in the Mississippian (*Conocardium*) and died out at the end of the Permian.

2.7 Brachiopods

Brachiopods (Phylum **Brachiopoda**) are marine organisms commonly called 'Lamp shells' on account of their resemblance to Roman oil-lamps. They first appeared in the Lower Cambrian and can still be found inhabiting deep water today. Like bivalves they consist of two valves or shells; however, they are different in that the two valves are unequal in size and are different in shape. Valves are hinged together along their posterior margin (Figure 62A–B). In advanced forms the hinge is complex with a tooth and socket arrangement. Shells (or valves) are **dorsal** and **ventral**, and the larger ventral valve often has an opening called the **pedicle foramen** through which a fleshy extension called a **pedicle** extends and attaches to the substrate (Figure 62C). Brachiopods were usually attached to the substrate, and some forms, particularly the Productids (Figure 63F) were stabilised in loose sediment by long spines that grew from the shell surface. Some other forms, such as the thin-valved *Chonetes*, simply lay unattached on the seabed where they probably lived in low-energy conditions (Figure 62E).

Internally brachiopods possessed a pair of muscles (adductor and diductor) that were used in the opening and closing of the valves (Figure 62D). When one set contracted the other relaxed, and the valves opened along the anterior margin. This produced a gap or gape along the **commissure** (junction) between the valves through which water carrying food could pass. Some brachiopods such as the

Rhynchonellids (Figure 63E) developed shells that had very **plicated** valves. This was an adaptation which allowed them to take in food particles of a larger diameter than those taxa that had a straight commissure. Brachiopods were **filter feeders** and used a spiral structure called a **lophophore** to extract the food. In some groups this was held on calcareous supports called spiral **brachidia** (Figure 64). In the Spiriferids (Figure 63B) these were elongated so that they became more efficient, and to accommodate them the shells expanded laterally into 'wings', which as a consequence resulted in the formation of a very long straight hinge line. In China, as well as in North Wales, many such brachiopods were known as 'Butterfly Shells'.

Brachiopods were common in the Lower Palaeozoic where they inhabited shallow-water environments. Some classic studies examined Silurian brachiopods in central Wales and found distinctive assemblages could be correlated with water depth. This allowed palaeontologists to determine the depths of deposition of sediments found elsewhere that contained similar assemblages.

At one time in the nineteenth and early twentieth centuries brachiopods and bryozoans were grouped together on account of both possessing a lophophore. In reality, this is morphologically quite different in each group. The classification of the Phylum Brachiopods is complex, and they have been conveniently subdivided into two classes based largely on

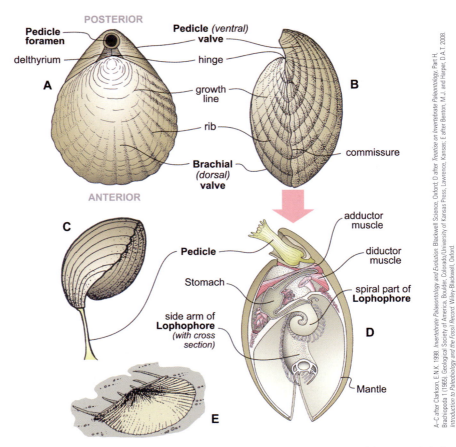

POSTERIOR

Pedicle foramen

delthyrium

Pedicle *(ventral)* valve

hinge

A

growth line

rib

B

commissure

Brachial *(dorsal)* **valve**

ANTERIOR

C

Pedicle

Stomach

side arm of **Lophophore** *(with cross section)*

adductor muscle

diductor muscle

spiral part of **Lophophore**

D

Mantle

E

A–C after Clarkson, E.N.K. 1998. *Invertebrate Palaeontology and Evolution*. Blackwell Science, Oxford; D after *Treatise on Invertebrate Paleontology*, Part H, Brachiopoda 1 (1965). Geological Society of America, Boulder, Colorado/University of Kansas Press, Lawrence, Kansas; E after Benton, M.J. and Harper, D.A.T. 2008. *Introduction to Paleobiology and the Fossil Record*. Wiley-Blackwell, Oxford.

Figure 62 Brachiopod morphology and various modes of life. A–B. Dorsal and lateral views of the external shell morphology of *Magellania flavescens*; C. Reconstruction of *Magellania* in life position; D. Cross-section through the living brachiopod *Terebratulina* showing general internal anatomy; E. Reconstruction of *Chonetes* in life position on the seabed.

the presence or absence of a hinge structure. The Class Inarticulata including *Lingula* have no hinge structures, and chitinous, phosphatic or calcareous shells. The valves are simply held together with muscles, the pedicle simply protrudes between the valves, the gut has an anus, and the lophophore is unsupported. Members of the Class Articulata are more advanced with

a hinge structure, calcareous shells, and a gut lacking an anus.

Brachiopods thrived and diversified through the Palaeozoic, and at their **acme** contained many Orders (Figure 63). Stocks were decimated at the Permian–Triassic extinction event, but some taxa survived. One remarkable survivor was *Lingula*, an inarticulate form that

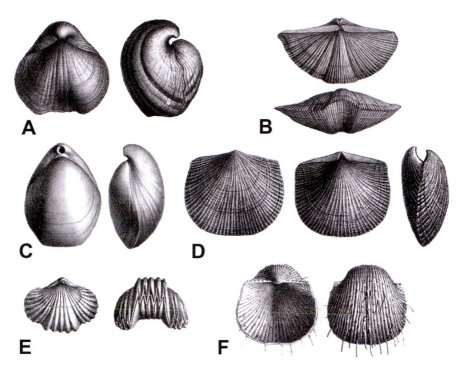

Figure 63 Brachiopod orders. A. Pentamerida; B. Spiriferida; C. Terebratulida; D. Orthida; E. Rhynchonellida; F. Productida. Lithographs from Thomas Davidson's monograph of the fossil brachiopods of Britain published between 1851 and 1886.

Figure 64 Spiriferid brachiopod showing the internal spiral brachidium.

After Clarkson, E.N.K. 1998. *Invertebrate Palaeontology and Evolution*. Blackwell Science, Oxford.

is found in modern oceans and shows little modification to its Lower Cambrian ancestors. During the Mesozoic bivalves began to dominate the shallow-water shelf areas previously inhabited by brachiopods, which were forced to live in deeper waters, away from the coast and from more successful competitors.

2.8 Echinodermata

The Phylum **Echinodermata** is so named because its members have an internal skeleton covered in bumps and spines so that the skin appears spiny. The Phylum includes a diverse group of organisms that has been divided into two major groups. The free-living **eleutherozoans** include the echinoids (sea urchins), holothurians (sea cucumbers), edrioasteroids, asteroids (star fishes) and ophiuroids (brittle stars), while the **pelmatozoans** are attached and comprise the crinoids (sea lilies), blastoids, primitive eocrinoids, and the rare cystoids that were confined to the Palaeozoic.

While appearing very diverse, all of these groups have a number of characteristics in common. Firstly, all of the groups show five-fold or pentameral symmetry. In the irregular sea urchins this is less obvious due to changes in the morphology of the **tests** brought about over millions of years adapting to new ecological settings. Secondly, all possess a **mesodermal** skeleton that is sandwiched between flesh. The skeleton comprises plates or **ossicles** that are loosely held together. Under high magnification it can be seen that the ossicles are very porous, but each is in fact a single crystal of calcite. Thirdly, all have a water vascular system (whose structure is best illustrated by reference to the sea urchins described subsequently). This series of internal canals give rise to the external tube feet that serve a multitude of roles.

What is known about the early fossil record of the echinoderms? The discovery of the Ediacaran faunas from the Flinders Range in South Australia in 1946 yielded a large number of oddities. Amongst these is *Arkarua* which is a disc-like fossil with a pattern of five rays on its upper surface. While it is unclear if it has echinoderm characteristics other than putative **pentameral** symmetry, it is possible that the echinoderms first appeared approximately 600 million years ago.

In the early Cambrian various groups evolved that bore echinoderm affinities. Amongst these were the helicoplacoids, which possessed a spiral arrangement of plates but did not have pentameral symmetry. In the stalked Eocrinoids there is little evidence of the water vascular system, and it would appear that they more closely resemble blastoids than

Figure 65 *Echinosphaerites* from the Ordovician of Estonia.

crinoids on account of the skeletal nature of their arms. The cystoid theca was spherical in shape, constructed on many hundreds of ossicles, and it was elevated into the water on a stem. While generally rarely preserved, good examples have been recorded from the Ordovician limestones of the Baltic region (Figure 65). Echinoderms have a complex early history; needless to say, by the earliest Ordovician most of the recognisable echinoderm groups had appeared.

2.8.1 Echinoids and Edrioasteroids

The **echinoids** or sea urchins make up the Class Echinoidea and are fascinating and complex organisms that had and continue to have a major impact on the ecosystems which they inhabit.

The first echinoids appeared in the Ordovician, and various groups have evolved since then. The differentiation and classification of these groups is largely based on the arrangement of the ambulacra and the modification of the body plan in some to enable a burrowing lifestyle. Echinoids have been classified into three subclasses, but herein it is considered that structure of echinoids is best explained by reference to the two now informal groupings: the regular echinoids (Figure 66A) and the irregular echinoids (Figure 66B).

All Palaeozoic and many of the post-Palaeozoic echinoids are regulars. They possess a characteristically spinose, spherical and radially symmetrical test, which is typically composed of ten vertical rows of plates, as displayed in the Mississippian genus *Palaechinus* and *Maccoya* (Figure 67A). Five rows of plates are perforate (**ambulacral** areas) and these alternate with five imperforate (**interambulacral** areas) rows (Figure 66A₁).

In life spines, which provide both a protective function and also raise the animal off the seabed, are carried on many of the plates and these can articulate on a rounded boss. In *Hemicidaris* from the Cretaceous and its modern descendant *Cidaris*, a Recent genus that lives in the Indian and Pacific Oceans, the spines are particularly impressive and can reach 10 cm in length (Figure 67B). In *Tylocidaris*, also from the Cretaceous, the spines are club-shaped. Many spines bear ridges, nodes or even backward-pointing barbs, which if stepped on can cause serious injury and even hospitalisation for the victim. In fossil echinoids it is rare to find specimens complete with spines.

Close examination of the ambulacra reveals a series of paired pores on each plate (Figure 66A₂). From each of these pores emerged **tube feet** and these are connected to the internal water-vascular system that allowed them to fill and became firm when needed. Tube feet play an important role in echinoids; they allow for the exchange of gas in respiration, can move food to the mouth, and are responsible for locomotion, even having the ability to right an over-turned animal.

The **peristome** or mouth is situated centrally on the ventral or lower surface of the echinoid (Figures 66A₂, B₂). This is an opening covered by small plates, beneath which is an apparatus called the **Aristotle's Lantern**. This muscular structure comprises five interlocking teeth that can grab food and pass it to the gut behind. The gut ends in the anus, which opens in the **periproct** on the dorsal or upper surface (Figures 66A₁, B₁). Surrounding the periproct is a series of ten plates that make up the apical system; the largest of these is the **madreporite** (Figure 66A₁) which resembles

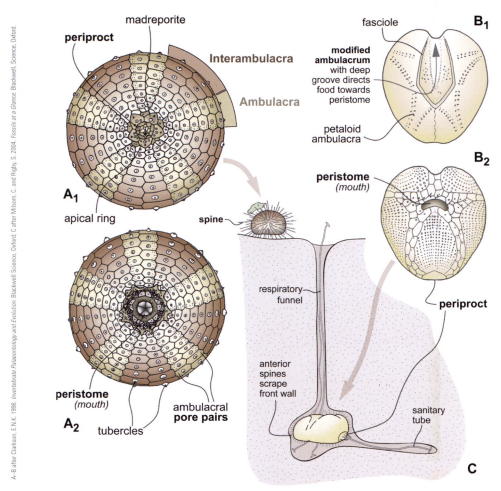

Figure 66 A. Morphology of *Echinus esculentus*, a living regular echinoid; A$_1$. Dorsal view; A$_2$. Ventral view with the central peristome containing the Aristotle's Lantern. B. Morphology of *Echinocardium cordatum*, a living irregular echinoid; B$_1$. Aboral (top) view; B$_2$. adoral (bottom) view; C. Both echinoids shown in life position: *Echinus* on the seabed and *Echinocardium cordatum* in a burrow in the sediment with long tube feet extended to the seawater and with a sanitary tube behind the test.

a sieve and allows water to enter the water-vascular system below. This is made up of a circular canal from which five lateral canals emerge, which lie beneath the ambulacral plates, and from which the tube feet develop.

The gonads of the echinoids are situated beneath the interambulacral plates.

Regular echinoids had an epifaunal habit, living on top of the sediment surface, whereas modern forms tend to prefer low-energy environments where they are less susceptible to damage. They are voracious predators that in some environments number many hundreds per square metre. They have the ability to move across the seabed, devouring and stripping it of organisms and leaving it often rather barren.

Irregular echinoids such as *Micraster*, on the other hand, first evolved in the Jurassic to facilitate a burrowing mode of life (Figure 66C). This presented a logistical problem in that the anus in regular echinoids is situated on the upper side of the test, and so burrowers would defecate on themselves. To solve this, the

Figure 67 A. *Maccoya sphaerica*, a regular echinoid from the Mississippian of Millicent, Co. Kildare, Ireland; B. Isolated spines from the regular Cretaceous echinoid *Cidaris*; C. Fossil irregular echinoid from the Cretaceous of Mali: Upper aboral surface with ambulacra and periproct (anus) on posterior (lowermost) side; D. As C: Lower adoral surface with bean-shaped mouth; E. *Scutella*, a sand dollar from the Oligocene of Wadihassan district in the Sahara Desert; F. *Cystaster stellatus*, an edrioasteroid encrusted with the runnerlike bryozoan *Corynotrypa* from the Ordovician of northern Kentucky, USA.

periproct containing the anus migrated to the side of the animal, and this zone of migration manifests itself in fossil and modern irregular echinoids in a deeply grooved ambulacrum that has resulted in a heart-shaped and bilaterally symmetrical test (Figures 66B, 67C–D). A long tube foot excavates a sanitary burrow behind the animal, into which waste materials are packed. Burrowing is facilitated by short anterior spines that excavate sediment, and much of this sediment is ingested and the contained organic matter digested. In order to respire, these infaunal animals extend a long tube foot up via a funnel through the sediment to the seawater.

By the early Tertiary a number of the irregular echinoids further evolved a lifestyle in which they were either not buried, where they adopted a **benthic** mode of life in high energy conditions, or else were shallow sediment burrowers. These were characterised by having tests with a flattened bottom (as in the Miocene genus *Clypeaster*) and further modification to a highly flattened test gave rise to the sand dollars such as *Scutella*, which is known from the Oligocene sediments of the Sahara Desert (Figure 67E). The flattened test is a modification to high-energy muddy environments, and notches in the side of the test act as a 'spoiler', which reduces the possibility of the echinoid being tipped over by water currents.

Edrioasteroids (Class Edrioasteroidea) have a fossil range that began in the Ediacaran 600 Ma ago if one accepts that *Arkarua* belongs to this group. Undoubted edrioasteroids such as *Cystaster stellatus* (Figure 67F) appeared in the Lower Cambrian, and they disappeared by the Permian. They consisted of a spherical theca made up of numerous plates, which in some genera sat on a short stem attached to the substrate. Other genera were thought to be free-living. In the centre of the upper dorsal surface was a mount covered by a series of plates. Radiating from the mouth were five sinuous ambulacra of small plates which were divided from each other by interambulacra of larger polygonal plates.

2.8.2 Crinoids and Blastoids

Members of the Phylum Echinodermata that are generally attached to the seabed by a stem are grouped into the Plematozoa and include the crinoids and the **blastoid**s.

Crinoids (Class Crinoidea) are popularly known as sea-lilies, but they are animals and not plants. They appeared in the middle Cambrian and are still found in modern oceans. Most are fixed to the seabed by means of a holdfast, but some rare forms may be free-living swimmers such as the comatulids or feather-stars (*Saccocoma* and *Antedon* (Figure 68C). The latter had a stalk early in life but lost it when it reached adulthood. When required it could attach itself temporarily to the sediment by means of cirri at the base of the calyx (Figure 68C). The attached crinoids consist of a stem that is made up of ossicles or **columnals** (resembling polo mints) stacked on top of each other and a **calyx** (cup) with five pinnate arms typified by the living crinoid *Ptilocrinus* (Figure 68A). The columnals display a range of outline shapes. All have a central opening or lumen that carries a thread of flesh down the stem which controls movement, and while the **lumen** is often circular in outline, in some genera it is pentalobate reflecting the **pentameral symmetry** of the group. Some columnals are grooved, and these fit into each other and

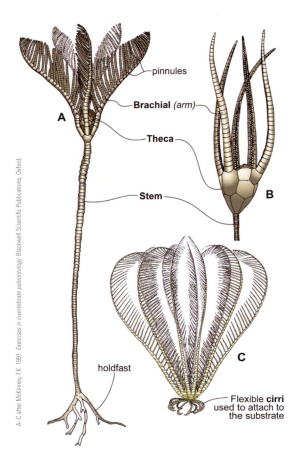

Figure 68 A. Generalised crinoid morphology showing a complete specimen of *Ptilocrinus pinnatus*, a living crinoid, with a holdfast, long stem, a theca and outspread brachials bearing fine pinnules; B. *Hybocrinus* with uniserial arms, from the Middle Ordovician of the USA; C. *Antedon*, a modern stalkless comatulid crinoid; D. *Dialutocrinus aculeatus*, a crinoid from the Mississippian of Hook Head, Co. Wexford, Ireland.

produce a rigid stem, while in others they have a central ridge which articulates with that of its neighbours above and below, and this allows the stem some movement. Some taxa are known from columnals only.

The calyx contains the soft tissue, and it is made up of a number of rigid plates arranged in rows from its base upwards. It is attached to the stem by means of five basal plates, and on top of these are found the radial plates to which five arms or **brachials** are attached. The top of the calyx is made up of a number of covering plates that form a dome or **tegmen**. The brachials may bear small alternate **pinnules** that give them a feathery appearance (Figure 68A), or these may be lacking in some taxa such as *Hybocrinus* from the Ordovician of the USA (Figure 68B). The arms spread out to form a filtration fan, and the pinnules increase its surface area for capturing food particles. In some taxa such as *Encrinus* from the Triassic the arms are divided a short distance along their length, and this also improves their efficiency.

Food is passed down a food-groove on the arms and enters the mouth. Crinoids do not have a stomach but digest food in an intestine. This ends at the anus, which opens out at the edge of the tegmen. Associated with the anus in some taxa are **coprofagous** gastropods.

Modern-day crinoids live in a range of water depths from shelf environments to deep oceanic settings. Stalkless crinoids generally live in shallow water, while stalked, attached crinoids are found below 100 m water depth. Some crinoids are **rheophyllic** (current-loving) and a number can twist their stems so that the filtration fan can be positioned to best exploit the water currents that carry food. Deeper-water crinoids generally are **rheophobic** and avoid currents, preferring to have a more passive lifestyle where food simply falls into the outstretched filtration fan. Palaeozoic crinoids were more conspicuous inhabitants of shallow-water shelf areas than are their modern counterparts. Studies have shown that a number of assemblages of crinoids arranged themselves into different tiers or levels in the water so that they were not in direct competition for food. The stems of crinoids often attract epiphytic organisms such as brachiopods, corals (*Emmonsia*), and encrusting bryozoans, including *Fistulipora* and *Meekoporella* from the Mississippian. These benefit from being elevated above the sediment–water interface where cleaner water and food are found.

Some crinoids unusually developed a pelagic mode of life: loboliths, which were buoy-like floating devices, carried Ordovician crinoids and *Pentacrinites* from the Jurassic attached themselves to floating logs.

Crinoids exhibit a considerable range in size: some microcrinoids were less than a centimetre in height (*Pygmaeocrinus*) while the largest formed thecae 30–40 cm in height. Some crinoid stems reached over 2 m in length.

Many limestones contain the remains of disarticulated crinoids, in particular the stem ossicles. On death the thin outer soft tissue decays and the plates fall apart. In rare examples complete crinoids have been preserved, such as those from the Mississippian of Crawfordsville, Indiana and Hook Head, Ireland (Figure 68B), and it is thought that these were buried rapidly following a storm event and before the skeleton could become scattered.

Blastoids (Class *Blastoidea*) resemble crinoids, as they consist of a stalk surmounted by a theca or cup, but there the similarities end. Arising from the calyx was a crown of **brachioles** which were used for food gathering (Figure 69). The theca, which measures between 5 and 30 mm in height, shows strong pentameral symmetry (Figure 69). The brachioles are developed on ambulacra, but in fossils these usually have become unattached and lost. These helped carry water and food towards the ambulacra, and food was passed along a food groove to the centrally-placed mouth. Along the edges of the ambulacra small pores carried water into the body, and this was passed over the **hydrospires**. Waste water was passed out through four small circular openings called spiracles situated around the mouth on the top of the calyx, and waste digestive products passed out through an adjacent anus. Gas exchange took place on hydrospires, which in more primitive blastoids were external, but in later genera such as *Orophocrinus* and *Pentremites* from the Mississippian (Figure 69) they were internal and highly invaginated, which increased the surface area for gas exchange.

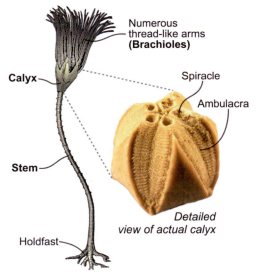

Numerous thread-like arms (Brachioles)

Calyx

Spiracle

Ambulacra

Stem

Holdfast

Detailed view of actual calyx

Figure 69 Blastoid morphology showing a reconstruction of a complete specimen of *Orophocrinus* from the Mississippian of the USA (left), together with *Pentremites tulipaformis*, a Late Mississippian blastoid from the Hardinsburg Formation of Christian County, Kentucky, USA. The spiracles on top of the theca are clearly visible, as are the ridged ambulacra.

Left after *Treatise on Invertebrate Paleontology*, Part S, Echinodermata 1(2) (1967). Geological Society of America, Boulder, Colorado/University of Kansas Press, Lawrence, Kansas.

The earliest blastoids are Silurian in age, and they reached a peak in diversity and abundance during the Carboniferous and Permian and disappeared at the end of the Palaeozoic. They inhabited similar shallow-water environments as did the crinoids. Usually blastoids are found disarticulated and often only the theca is found preserved complete while the stem and delicate brachioles become scattered.

2.8.3 Starfish, Brittlestars and Homalozoans

Starfish (Class ***Asteroidea***) are amongst the most easily recognisable of marine modern animals with nearly 2000 species described. They have a flattened test extended into arms, which are characteristically five in number, but may be more, in multiples of five. The under-surface of the arm bears tube feet, which in modern asteroids are used for opening molluscs, the usual food for this group. Some modern species, such as the Crown-of-thorns Starfish, are predators on tropical corals and can do serious and lasting damage to fragile ecosystems. In feeding the tube feet pass food to a food-groove situated on the underside of the arms that leads to the centrally placed mouth. The anus is situated on the **adapical** or upper surface close to the madreporite, which serves the same function as in echinoids, replenishing the water vascular system. Tube feet also help the animal to move and reorientate itself should it become overturned, and they are also involved in respiration.

The earliest starfish are Ordovician in age, but the fossil record of starfish and their close relatives the brittlestars is rare, as they are composed of loosely linked plates that are covered by a thin layer of flesh. These ossicles are usually granular, calcareous, of regular shape, and may be associated in rows. Exceptionally preserved Jurassic starfish are known from the Solnhofen limestone of Bavaria in Germany, while *Pentasteria orion* from Yorkshire is less well preserved (Figure 70A).

Starfish live in shallow-water environments and are active predators. When at rest they can leave imprints in the sediment, and these are occasionally preserved as the trace fossil *Asteriacites*, examples of which have been recorded from the Ordovician onwards of the USA and Europe.

Brittlestars (Class ***Ophiuroidea***) have a central disc that contains all the body organs, from which are derived long, thin and flexible arms. This typical arrangement is best

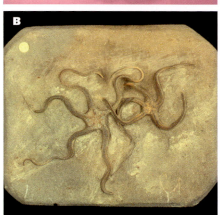

Figure 70 A. *Pentasteria orion* from the Jurassic of Kelloway Rock, Newton Dale, Pickering, Yorkshire, England; B. The brittlestars *Ophioderma egertoni* from the Lower Jurassic of Lyme Regis, southern England; C. *Dendrocystoides*, an Ordovician homalozoan, a group that is variously placed in the Echinodermata or the chordates.

illustrated by *Ophioderma egertoni* from the Lower Jurassic of Lyme Regis, southern England (Figure 70B). Brittlestars appeared in the Ordovician, and today they live in deep water, where they are predominantly suspension feeders. As their arms allow considerable movement in many orientations, they have utilised them for gathering tiny food particles.

Homalozoans. These are highly unusual animals that have been called homalozoans, 'carpoids' or calcichordates. The animal is covered by plates, which suggests affinities with the Echinoderms, but it lacks pentameral symmetry. Consisting of a long tail and an upper 'head', the latter contains a single arm or brachiole, a hydropore for water and food intake, gill slits and an anus. In some forms such as *Dendrocystites* from the Ordovician (Figure 70C) some of these features may be absent or hard to see. It is difficult to comprehend how this animal lived, but it may have used the tail for locomotion. Some palaeontologists have argued that homalozoans show close affinities to the chordates. They ranged from the Cambrian to the Pennsylvanian.

2.9 Arthropods

Arthropods are a morphologically diverse group that includes Trilobites, King Crabs, Ostracods and other crustaceans, and Insects. Until the 1980s these were classified together into a single phylum and divided into lower ranking taxonomic groups, but more recently many scientists consider the arthropods to be a **polyphyletic** grouping that comprises a number of phyla. The salient characteristics of four phyla are described in later sections.

Arthropods are so named from the Greek for 'jointed foot' and all have a bilateral body plan with a hardened outer skeleton, or **exoskeleton**, composed of chitin and proteins, which they need to shed or moult in order to grow. The chitin may be additionally hardened or **sclerotised** by the addition of calcium carbonate or calcium phosphate. The body is segmented, often into distinct regions, and they possess jointed limbs that may be modified to fulfil a variety of functions, including for catching prey, walking and swimming.

Arthropods were and remain important both in the marine realm and in the terrestrial realm. They have a long geological history extending back over 600 million years. Many bizarre forms were discovered for the first and only time in the Middle Cambrian deposits of the Burgess Shale from British Columbia, Canada. Over twenty different arthropods have been reported that make up approximately 40% of the biota, and some still perplex palaeontologists, who have difficulty deciding on their correct taxonomic placements or modes of life on account of their odd body arrangements.

2.9.1 Trilobites

Trilobites (Phylum **Trilobita**) are an extinct group of marine arthropods that occupied an important position in ecosystems in Lower Palaeozoic times.

Their body is covered by a hard chitinous–calcareous shell that is divided into three longitudinal lobes, hence their name. The central axial lobe and the two outer lateral lobes are separated from each other by two axial furrows, and the body is also divided into a **cephalon** (head shield), **thorax** and **pygidium** (tail) (Figures 71A, 72B). The cephalon is fused, but shows signs of former segmentation. Where eyes are present they are compound, made up of many lenses (Figure 71B). The **glabella** is an inflated bump, under which the stomach is situated. On the underside of the glabella is found the hypostome, which is a skeletal plate that supports the mouth and stomach. It arises from the doublure, that is the turned-in lower margin of the cephalon (Figure 71A). The majority of trilobites possess a **facial suture**, which is a line of weakness in the shell crossing the cheek. Four patterns of suture are known, and these have been utilised by palaeontologists as a means of classifying the animals. These are described as **proparian**, **opisthoparian**, or **gonatoparian,** according to the position of the suture. All start at the anterior margin, run through the

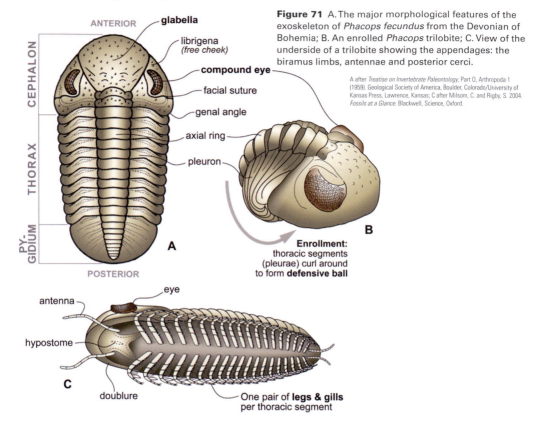

ANTERIOR — **glabella**

librigena
(free cheek)

compound eye

facial suture

genal angle

axial ring

pleuron

CEPHALON

THORAX

PY-GIDIUM

A

POSTERIOR

Figure 71 A. The major morphological features of the exoskeleton of *Phacops fecundus* from the Devonian of Bohemia; B. An enrolled *Phacops* trilobite; C. View of the underside of a trilobite showing the appendages: the biramus limbs, antennae and posterior cerci.

A after *Treatise on Invertebrate Paleontology*, Part O, Arthropoda 1 (1959). Geological Society of America, Boulder, Colorado/University of Kansas Press, Lawrence, Kansas; C after Milsom, C. and Rigby, S. 2004. *Fossils at a Glance*. Blackwell, Science, Oxford.

B

Enrollment:
thoracic segments
(pleurae) curl around
to form **defensive ball**

antenna

eye

hypostome

C

doublure

One pair of **legs & gills**
per thoracic segment

eye, and in the case of the proparian style, end on the lateral margin of the cephalon (Figure 71). Opisthoparian sutures terminate along the posterior margin of the cephalon, while the gonatoparian sutures emerge at the **genal angle**. Their function was to aid **ecdysis** or moulting. The thorax has a variable number of segments, and the pygidium is made up of a variable number of segments that have fused together. In some genera such as *Isotelus* the pygidium is smooth and the position of the former segments cannot be discerned (Figure 72D) whereas in other taxa pseudosegmentation can be seen. The relationship in size

between the cephalon and the pygidium is also used in classification, and four states are known: Micropygous (pygidium tiny compared with cephalon, typical of Middle Cambrian taxa); Heteropygous (pygidium slightly smaller than cephalon); Isopygous (pygidium equal in size to cephalon); and Macropygous (pygidium larger than cephalon). Some trilobites, such as *Calymene* and *Phacops*, were able to enroll or roll up during times of attack or when buried, and many are found preserved in this state (Figure 71B).

Preservation of soft tissues in trilobites is rare, but it does occur, as in specimens

Figure 72 A. Exceptional preservation in pyrite of the appendages of *Triarthrus eatoni* from Beecher's Trilobite Bed, Ordovician of Oneida County, New York, USA; B. *Ptychoparia striata* from the Cambrian of Ginetz, Bohemia, Czech Republic; C. Small agnostid pelagic trilobites from the Cambrian of Stemmstad, south of Oslo, Norway; D. *Isotelus gigas* a smooth isopygous trilobite from the Ordovician of the Trenton area, New York, USA.

recovered from the Middle Cambrian of the Burgess Shale or from the Devonian of upstate New York (Figure 72A), and these specimens have yielded a great deal of useful biological information. The head bears a pair of jointed antennae and four pairs of two-branched jointed **appendages** on the ventral side. On the ventral underside of the thorax and pygidium each segment bears a pair of two-branched or **biramous** appendages (Figure 71C). One portion comprised hollow jointed limbs that were used for walking or swimming, while the other contained a filament that may have served as a gill for respiration. Beneath the pygidium are found two posterior antennae called cerci.

Trilobites reproduced sexually by laying eggs. In some trilobites it is thought that they could hold larvae in **brood chambers** and release them into the water once they were sufficiently mature.

In order to grow, trilobites cracked their hard exoskeleton along the facial suture and the soft body crawled out during ecdysis. The shell was discarded, and where preserved these have left a record of their developmental stages. Growth occurred in increments; the earliest phase is called the **protaspid** period when the animal was a single, simple segment. The following **meraspid** period was reached when thoracic segments were added, one at a time per meraspid moult. Finally the adult **holaspid** phase was reached when the full complement of thoracic segments had been reached. Further moulting increased the size of the adult, which ranged in size from less than 5 mm to over 70 cm. The largest trilobite found is a species appropriately named *Isotelus rex* from Churchill, Manitoba in Canada, which reached 72 cm long. Preservation of complete trilobites probably occurred when they were rapidly buried. Isolated fragments represent moults that have broken up during ecdysis or by water current action. Cephala and pygidia rather than thoracic segments tend to be more frequently fossilised, as they are robust.

Trilobites lived in a range of ecological environments from shallow to deep water. They were burrowers living in the sediment (infaunal) or grazers living on the sediment surface (**benthic**), some were swimmers while others, such as Cambrian agnostid trilobites (Figure 72C), were so small that they floated in the surface waters. Interpretation of their mode of life is based on examination of both the morphology of their exoskeleton and the sediment in which they are preserved. Many spinose trilobites may have been floaters, the spines increasing the area of the body but being lightweight. Swimming trilobites often had a wide axial lobe in which was situated a muscular **coxa** that moved the limbs utilised for swimming. Eyes on some swimmers were huge and held above and to the side of the body, which allowed them a wide field of view.

Some are found mainly in shallow sea deposits, where they probably crawled over the sea bottom in search of food, or borrowed in the sediments. Some occupied deep water environments beyond the photic zone – many of these forms are blind and bear sensory fringes along their frontal anterior margin (*Harpes*, *Trinucleus*). As mentioned above, there is some indication that some trilobite forms may have been swimmers or floaters rather than crawlers. Trilobites preserved in many limestones were shallow-water forms, while those found in shales were either deepwater forms or their discarded exoskeletons were washed into deepwater areas from the shallows.

Trilobites first appeared in the Early Cambrian. They radiated rapidly through the Middle Cambrian and attained maximum development in Late Cambrian time. They were still very numerous in the Ordovician but began to decline during the Silurian. Still fairly abundant in the Devonian Period, they diminished rapidly during the Late Palaeozoic and completely disappeared by the close of the Permian.

2.9.2 Crustaceans

Crustaceans (Phylum **Crustacea**) are highly successful and include the crabs, lobsters and shrimps (Class **Malacostraca**), ostracods (Class **Ostracoda**), and barnacles (Class **Cirripedia**).

Malacostracans

Crabs are now conspicuous inhabitants of the marine system that are easily identified by virtue of their distinctive carapace, eyes raised on stalks, and large pincers that are just one of five pairs of appendage attached to the head. They have a short tail and the **abdomen** is completely obscured beneath the thorax when viewed from above. Some genera exhibit sexual dimorphism where the males are generally larger than the females. Modern crabs range in width from a few millimetres to several metres.

In the fossil record the oldest crabs may be *Imocaris* described from the Pennsylvanian of Arkansas in the USA, if one accepts that this taxon is a true crab. Jurassic crabs are well known, and there was a considerable radiation of the group in the Cretaceous. Often crabs are preserved in a disarticulated fashion with the loss of the pincers and other appendages, so that only the robust carapace is seen. Occasionally epibionts encrusting the carapace are also preserved. These sessile organisms, which included bryozoans, reaped the benefit of being continually moved into cleaner water by the host as it scuttled along sidewise across the sediment surface.

Shrimps and the larger lobsters, too, are readily identified on account of their elongate body plan with a long abdomen. The earliest crustaceans were the Phyllocarids, which originated in the Silurian. The Granton Shrimp Bed near Edinburgh, and other localities at the same stratigraphical level in Scotland, have yielded many beautiful examples of Namurian (Mississippian) eomalacostracean shrimps, perhaps the most striking of which is *Tealliocaris woodwardi* (Figure 73A) with its long antennae, first named in 1908 by Benjamin Peach. Other taxa include *Crangopsis* and *Palaemysis*. This is the same horizon that provided the first conodont animals. Shrimps and lobsters diversified during the Mesozoic and well-known, exceptional Jurassic examples such as *Cycleryon* are found in the Solnhofen Limestone (Figure 73B).

Figure 73
A. *Tealliocaris woodwardi*, a late Pennsylvanian malacostracan from the Granton Shrimp Bed, near Edinburgh, Scotland. B. *Cycleryon*, a decapod crustacean from the Jurassic Solnhofen Limestone, Germany.

Ostracods

Ostracods have bodies that appear unsegmented. The head, thorax and appendages are ill-defined and are contained in a calcite carapace, which in almost all taxa measures only up to 1.5 mm in width, but exceptionally in the Silurian genus *Leperditia* can reach 2 cm wide (Figure 74). The two valves are linked dorsally by an elastic ligament and a hinge, and the body hangs inside the carapace like a sac. There are seven pairs of appendages that serve as sense organs and for the capture and mastication of food, for locomotion and for cleaning the internal cavity. Ostracods possess a digestive system, complex genital organs, a central nervous system and a median eye within the carapace. Some **carapaces** are very thin and translucent, and so the animal could probably detect light, albeit rather poorly.

Ostracods have a long geological record, first appearing in the Cambrian and ranging to the present. At least three major groups have survived for over 500 million years. They lived in a variety of habitats, from deep marine settings, to pelagic swimmers, to rivers, lakes and swamps. On account of their size and on account of the many thousands of genera that have evolved and disappeared over time, they have been utilised widely by the petroleum industry for biostratigraphy, and correlating geological horizons, particularly those of the Mesozoic and Cenozoic.

Cirripedes

Cirripedes are barnacles, named from the Greek, meaning 'curled foot'. Most, such as the Acorn Barnacle, are sessile, and it will surprise many people that these small organisms, which are common in intertidal areas along the shoreline where they are attached to rocks, are in fact crustaceans. Six plates combine to make up a pyramidal-shaped shell which cements itself to the substrate such as rock or shells (Figure 60C). The animal can open two plates at its apex, and it agitates its legs to gather microscopic **plankton**. Some other species, such as the Goose Barnacles, attach themselves to the substrate by means of a fleshy foot similar to the pedicle in brachiopods. Cirripedes have a long but poorly documented geological record. Pedunculate species have been reported from the Middle Cambrian Burgess Shale, from the Silurian of Estonia, and from the Cretaceous of North America, the UK and the Caucasus. While we remember Charles Darwin for his work on the H.M.S. *Beagle* and on evolution, he also was a student of barnacles, and published a significant multi-part monograph on fossil pedunculate cirripedes in the 1850s.

Figure 74 *Leperditia*, an unusually large ostracod 20 mm wide from the Silurian of the island of Gotland, Sweden.

2.9.3 Chelicerates

Chelicerates (Phylum **Chelicerata**) include the horseshoe crabs and eurypterids (Class Merostomata) as well as the spiders (Class Arachnida). They are characterised by having a fused **prosoma** (head and thorax), a segmented **opisthosoma** (abdomen) and a pair of pincers called **chelicerae** attached to the prosoma.

The horseshoe crabs belong to the subclass Xiphosura and include the latter-day *Limulus*, and this provides a great deal of valuable information about its fossil counterparts, and to a lesser extent the trilobites. Although possessing a hard outer cuticle, the fossil record of xiphosurids is generally poor. They first appeared in the Cambrian, have also been recorded in the Pennsylvanian (*Belinurus*), and at least three genera have been reported from the Mesozoic, including important specimens of *Mesolimulus* from the Solnhofen lagerstätten of Germany (Figure 75A) where unusually the walking trail (trace fossil) of the animal ends with the actual body fossil of the animal itself. All xiphosurids follow a similar body plan with a large prominent prosoma

on which small eyes are placed laterally, up to twelve abdominal segments, and a posterior telson (or spine). Beneath the carapace is a series of five paired jointed limbs which can be used for walking or for swimming. *Limulus* inhabits shallow-water environments where it is an active predator at night. By day it buries itself just beneath the sediment surface. It can walk on land for short periods of time, and unusually, swims upside down.

The Eurypterids (sea scorpions) are contained in another subclass, and these became extinct at the end of the Permian. These animals could grow up to 2 m in length and were the largest arthropods in the fossil record. They had a large prosoma with a round anterior margin, an opisthosoma of at least twelve segments that was divided into two parts, front and back, and a posterior telson. The chelicerae numbered five pairs. These were used to capture prey, and also as walking limbs, but in *Ptergyotus* (Figure 75B) and some related taxa the fifth chelicera was enlarged

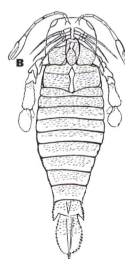

Figure 75 A. The underside of the carapace of *Mesolimulus walchi*, a horseshoe crab from the Jurassic Solnhofen Limestone, Germany; B. *Pterygotus anglicus*, a eurypterid from the Devonian of Scotland.

and its termination formed into a paddle used for swimming.

The arachnids include spiders and mites. The former are a fascinating group and surprisingly, given that most are terrestrial, they have a good fossil record that originated in the Silurian. Today there are over 100,000 species of spider. A number of fossil taxa are known from the Eocene Amber of the Baltic, which was derived from pine resin, while others have been preserved in sideritic nodules in the Pennsylvanian of Lancashire in England (*Phalangiotarbus subovalis* (Figure 76A)) and Mazon Creek, Illinois in the USA. While originally classified as spiders, they are also considered to have had affinities with the harvestmen. Recent studies have revealed fossilised silk used to make webs as well as the spinnerets that produced it. Mites are microscopic and have a long fossil record extending back some 390 million years.

2.9.4 Insects

Members of the Phylum **Uniramia** include myriapods such as centipedes and millipedes and also the insects, which are the most successful of all terrestrial animals. These organisms have a **uniramous** limb (unlike the trilobites) and use the tip of a modified limb as a jaw. They exhibit bilateral symmetry, and most are terrestrial.

Myriapods have a long geological record extending back into the Silurian, and while most are now rather small, the Pennsylvanian millipede *Arthropleura* reached over 3 m in length (Figure 76C). It remains the largest known terrestrial invertebrate.

Insects are exceedingly diverse and include flies, wasps, ants, crickets, weevils, earwigs and dragonflies. The group first evolved during

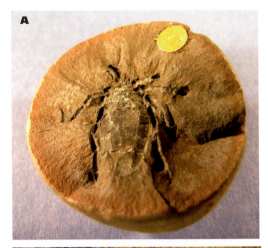

Figure 76 A. *Phalangiotarbus subovalis*, a spider from the Coal Measures (Pennsylvanian) of Burnley, Lancashire, England. The sticker indicates that this is a type specimen; B. *Libellulium*, a dragonfly from the Jurassic Solnhofen Limestone, Germany; C. Trackway made by the giant late Mississippian myriapod millipede *Arthropleura*, Isle of Arran, Scotland.

the Devonian. They have three pairs of jointed legs, compound eyes, and antennae, and some have a pair or two pairs of wings. Today over 10 million insect species are known. Most insects maintain a solitary existence, but some, such as wasps and bees, live in colonies, and as evidenced by the discovery of fossilised wasps' nests in the USA, have done so since the late Triassic, and perhaps earlier. Among the most impressive fossil insects are the dragonflies (Figure 76B) whose wingspan reached 80 cm. These reached their **acme** 300 million years ago during the Pennsylvanian when they would have flitted over the coal-forming swamps of Europe and North America.

Geologically insects are important, as some **coevolved** with the flowering plants, the angiosperms, during the Cretaceous, leading in some cases to strong mutualism when each brought benefits to the other. The insects often gathered nutrients while the plants benefited from increased pollination.

2.10 Graptolites

Graptolites (Class Graptolithina) are a group of extinct colonial organisms (Figures 77, 78) that were common in Lower Palaeozoic seas where most, but not all, drifted in surface waters, having a planktonic lifestyle. Morphologically their colony form is quite varied and this, together with the occurrence of numerous taxa, has resulted in graptolites being useful biostratigraphically as zone fossils. In the Ordovician and Silurian many subdivisions are based on graptolites. They first appeared in the Upper Cambrian and became extinct at the end of the Pennsylvanian.

Graptolites are most often found in black shale, were they resemble silver to grey-green coloured pencil marks (Figure 31). This gives them their name from the Greek *graphein*, to write, and *lithos*, stone. When the animals died the skeleton (or **rhabdosome**) slowly sank through the water column, finally resting on the seabed. In shallow-water deposits the rhabdosomes are rarely found, as they became obscured by bioturbation and other fossils. In very deep water where conditions are anaerobic, lacking oxygen, other animal forms are lacking and the rhabdosomes survive. By and large these are preserved flattened, and the original skeletal material, which is chitinous, may be altered to minerals such as chlorite. In rare cases the colonies are preserved in iron pyrites together with their original three-dimensional shape.

Exceptional preservation of the original skeletal material is known from specimens extracted by acid digestion of limestones (Figure 78B–D) and this has provided invaluable information pertaining to the biological affinities of the group.

Taxonomists have subdivided the Class Graptolithina into several orders, of which two, the Order **Dendroidea** and the Order **Graptoloidea**, are the most important. The latter appeared in the Middle Cambrian and formed bush-like colonies up to 400 mm in size. The **stipes** (branches) of the colony possessed two internal chambers called thecae, and were held together by crossbars called **dissepiments**. A typical dendroid graptolite such as *Dictyonema* was sessile, attached to the seabed by a holdfast, and so benthic (Figure 78A). Growing from the holdfast was the first theca called an autotheca, and this housed the individual animal. From this additional autotheca were added by budding. Associated with the **autotheca** were smaller cylindrical **bitheca** and a **stolon** than ran through the whole colony. The function of the bitheca remains unanswered.

It is thought that the members of the Graptoloidea evolved from the Dendroidea during the early Ordovician, when there was a transition from a benthic to pelagic mode of life. This involved the transformation of the holdfast into a thin pointed nema which may have been attached to a float. Following this transition the morphology of the rhabdosome changed markedly. The bitheca were lost and only the autotheca remained, and the number

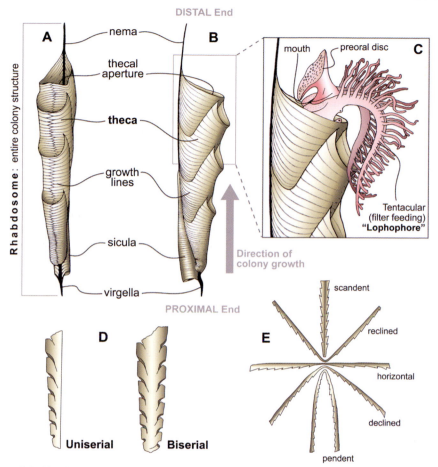

Figure 77 A–B. Morphological features of a generalised *Monograptus* graptolite; C. Hypothetical reconstruction of a graptolite zooid. A polypide of the modern pterobranch *Rhabdopleura* is shown inserted into a theca of a monograptid rhabsosome; D. Uniserial and biserial rhabdosomes; E. Composite diagram showing the various arrangement of biserial rhabdosomes.

C after McKinney, F.K. 1991. *Exercises in invertebrate paleontology*. Blackwell Scientific Publications, Oxford; D after Clarkson, E.N.K. 1998. *Invertebrate Palaeontology and Evolution*. Blackwell Science, Oxford.

of stipes were reduced first to eight, then four, two and finally one in Silurian and Devonian monograptids. The overall plan of grapto- loidea morphology is best exhibited in *Mon- ograptus* (Figures 77A–B, 78B). The earliest part of the colony is the sicula, which is a downward-pointing cone, from which thecae

successively bud in the opposite direction and deviate away from the dorsal nema. In some genera stipes grow back to back in a biserial pattern (Figure 77D). The rhabdosomes also appear in different orientations so that some are horizontal and some point downwards from the nema, as in the pendent tuning-fork

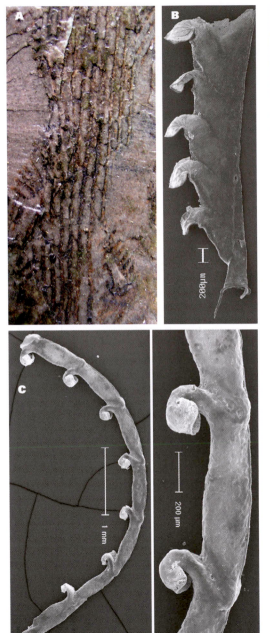

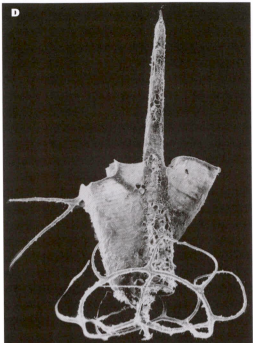

Figure 78 A. *Dictyonema*, a Cambrian to Ordovician bushy dendroid graptolite; B. *'Monograptus' birchensis* from the uppermost Pridoli, Silurian of Arctic Canada; C. *Streptograptus galeus* from the Upper Homerian of Arctic Canada; D. Specimen of *Pseudorthograptus inopinatus* from the Llandovery, Silurian showed the periderm rhabdosome wall consisting of successive half-rings.

graptolite *Didymograptus*, while others show an opposite plan as in *Leptograptus* (Figure 77E). The thecae developed a variety of morphologies, as did their terminations. Most thecae were broadly cylindrical, but their angle of deviation from the nema varied from high to low. They may have simple openings, or form into hooks (*Streptograptus* (Figure 78C)), lobes (*Lobograptus*) or even side lappets.

How did graptolites live, and could they swim? This vexed question is difficult to answer. Some researchers have examined gross rhabdosome shapes and have argued that spiral-shaped forms could move up and down in the water column, and perhaps some locomotion could be effected by the animals themselves. Other workers have argued that the rhabdosomes are so light that they would simply have floated, and were carried along in no particular pattern except those caused by the directions of the oceanic currents.

For a considerable time the biological affinities of graptolites puzzled palaeontologists, and it was only in the 1960s, with the advent of **Scanning Electron Microscopes**, that clues began to be pieced together. Detailed research on 3-D specimens showed that the **rhabdosome** wall, or periderm, was a twin-layered organic structure. The inner **fusellar wall** consisted of successive half-rings which interlocked with an opposing set to form a tube (as in *Pseudorthograptus* (Figure 78D)). Strengthening this arrangement was an outer **cortical layer** composed of cortical bandages that resembled strips of material used in plaster casts on broken limbs. Biologists were aware of modern-day organisms with a similar wall structure: this was *Rhabdopleura*, a **pterobranch** which resides in deep oceanic environments. What was fascinating, too, was the fact that *Rhabdopleura* is also known from the Cambrian, and so has a fossil record over 500 million years old. *Rhabdopleura* possess a pair of tentacles used for feeding, a stomach and intestine, and are held in cylindrical tubes by a disc-like structure. They are capable of moving outside the tube when feeding, but remain attached to the colony by a stolon, as are graptolites. We can gain some idea of what a graptoloid zooid looks like by using *Rhabdopleura* as a proxy (Figure 77C). Should this relationship be correct, then one must assume that graptolites and pterobranchs shared a common ancestor some time in the early Cambrian. The relationship is significant, as pterobranchs are **hemichordates**, a group related to vertebrates, and so too were graptolites.

In 1879 Charles Lapworth of the University of Birmingham recognised that some Palaeozoic graptolites that he had described from Scotland formed distinctive assemblages that differed from those known from Cambrian and Silurian rocks in Britain. As a result he coined the term 'Ordovician' for a geological period comprising the successions containing these assemblages that occurred between the Cambrian and Silurian. In so doing, he brought to an end the bickering that had seen Adam Sedgwick and Roderick Murchison and later others debate the true extent of the Cambrian and Silurian and the position of the boundary between the two.

2.11 Conodonts

Conodonts (Phylum Conodonta) are tooth-like structures that range in size from 0.2 mm to 6 mm. The earliest conodonts are Cambrian, and they disappeared from the fossil record in the Mid-Triassic, having survived the end-Permian extinction event. They are found in a variety of **lithologies**, which demonstrates that they were nektonic animals swimming in open marine waters.

Conodonts were first described in the mid-1800s by Christian Pander, a German palaeontologist, who coined the name 'cono' (conical) 'dont' (teeth). They perplexed many scientists over many years – Pander considered them to be the remains of fish, Roderick Murchison argued in 1854 that they were terminal portions of the segments of trilobites, while the English anatomist Richard Owen identified them as spines and hooklets from molluscs or annelid worms. It wasn't until the 1980s that the actual biological affinities of conodonts were finally revealed. Conodonts are composed of the mineral Francolite, a calcium fluorite apatite which is a phosphate mineral, and this unusual chemical composition distinguishes them from all other microfossils.

Morphologically, conodonts fall into three groups, and each conodont is referred to as an element: (1) Simple cones, with the appearance of a tooth or a hook. These have a base with a hollow basal cavity that sits on the jaw, and a long pointed cusp that may be straight or deflected to the anterior or posterior; (2) Ramiform elements, with a principal conical cusp flanked laterally by blades from which pointed **denticles** are developed (Figure 79A); and (3) Platform elements, with a blade partially bordered by two lateral extensions covered with nodules and ridges, with a denticulated **carina** positioned centrally (Figure 79B). The simple cones were found in animals that became extinct at the end of the Devonian, whereas the other two element types persisted through to the early Mesozoic.

Within a sample many different conodont elements of a wonderful variety of forms and shapes can be found, and consequently it was difficult for early taxonomists to decide how to name all of them. What was known was that conodonts exhibited symmetry, and left-hand and right-hand forms could be matched to each other; but they did not know, as Pander worried, if they came from one individual animal or from a number of animals. The solution to this taxonomic conundrum was to give each morphological type its own generic name – this is called 'Form Taxonomy' and each taxonomic rank a '**Form Genus**' or '**Form Species**' as appropriate. Later in the 1930s some fortuitous finds in the Mississippian limestones of Illinois threw up assemblages of different conodont types that were found displaying some arrangement on bedding planes. It was quickly realised that these conodonts made up a '**conodont apparatus**' that reflected the arrangement of the conodonts in life position. It showed that the Ramiform

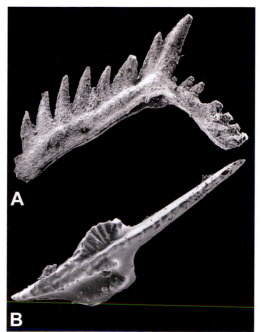

Figure 79 Conodonts. A. *Idioprioniodus* cf. *healdi*, a ramiform element; B. *Gnathodus texanus*, a platform element [both A and B are from the Mississippian of South Wales]; C. Head of a conodont animal from late Mississippian Granton Shrimp Bed, near Edinburgh, Scotland.

elements were grouped together slightly in front or behind pairs of Platform elements.

However, the precise function, arrangement, and biological affinities of the conodonts were not unravelled until 1983, when the first complete fossil conodont animal was discovered in the Mississippian Granton Shrimp Bed close to Edinburgh, Scotland (Figure 79C). These rocks contain an exceptional suite of finely preserved crustaceans, and loitering amongst some shrimps was a tiny worm-like animal that had an assemblage of conodonts preserved at one end. This was, without doubt, a true Conodont Animal that was later named *Clydagnathus windsorensis*. It was 8 mm in length and possessed clearly

preserved muscle along its trunk, a post-anal tail, and two expanded lobes at its anterior end, which were interpreted as eye sockets. Today ten conodont animals have been recovered, with the largest known from the Soom Shale of South Africa, where the animal reached 200 mm in length.

Conodont animals have many of the characteristics that make placement in the Chordates seem reasonable. They have eye structures, a post-anal tail, and a **notochord**. For the moment many researchers have retained the Phylum Conodonta to accommodate these animals. The closest modern analogue are the Arrow Worms or Glass Worms, which are small but voracious planktonic carnivores.

Research on the conodont apparatus and studies of individual elements under the Scanning Electron Microscope have suggested that conodont animals were major predators. Wear on denticles shows that they were used for biting, and it is probably that the elongated ramiform elements were situated at the front of the mouth in the animal. There they could hold onto prey, which was then passed down into the **buccal cavity** where the platform elements would have sliced up and ground down the prey into easily swallowed pieces.

From many different rock types conodonts can be extracted in large numbers, and they have a number of useful geological applications. They are of critical importance in the study of biostratigraphy, and can be used in the **correlation** of different rock successions from area to area. This is because conodont morphospecies evolved rather rapidly, and so the time ranges of many species can be short, and recognition of these short-ranged forms in a particular rock can tightly constrain its age. Use of these has led to the erection of conodont biozones indicative of approximately 3 million years each for various parts of the fossil record. The Mississippian of western Europe is often subdivided on the basis of conodonts. In some cases rocks can be correlated on the basis of an assemblage of different conodonts rather than relying on the presence of just one species, and sometimes, as evolutionary lineages can be recognised in conodonts, the disappearance of a species or the sudden appearance of another can be used to tie down a particular time.

In the petroleum industry it is important not only to know the age of rocks that may contain oil and gas, but also the temperature that these rocks reached after burial. If too hot, then hydrocarbons may be lost. If too cool, then hydrocarbons probably never formed in them. Conodont studies provide a basis by which these palaeotemperatures can be determined, and the rank of the rocks measured. As they are subjected to increased temperatures, conodonts change colour, going from pale to dark. A simple examination of the colour provides an estimate of the temperature reached by sedimentary rocks.

Without doubt, conodonts are one of the most fascinating fossil groups known, and students of the group continue to study them using any new research methods available to them. In recent years analysis of nitrogen isotopes preserved in conodont elements is now suggesting where the animals lived, and indeed what food types they ate.

2.12 Fishes

In the fossils and modern fauna there are more species of fish than the other vertebrates put together. Fish are characterised by the possession of gills, a vertebral column and a single-looped circulatory system. They are covered by **scales**, have fins for propulsion, and more derived forms have a swim-bladder which regulates buoyancy and a sensory system – the lateral line, aligned along the body.

The bulk of fossil fish are represented by teeth, scales or spines, and whole specimens are rather rare. In recent decades research on ichthyoliths (fish microvertebrates) have also added to the understanding of Lower Palaeozoic fish faunas and phylogenies.

The oldest fish were the jawless **ostracoderms** that first appeared in the Cambrian about 470 Ma. Specimens have been recovered from the Georgina Basin in Australia. The jawless fish essentially resembled the modern-day lampreys and hagfishes. By the Silurian and Devonian many of these developed heavy armour that covered the head and front portion of the body; this group are the heterostraci fishes, which include over 300 species in genera such as *Anglaspis and Erriv-aspis* (Figure 80A, B). Living alongside them at the same time were other less well-armoured ostracoderm fishes including *Rhyncholepis* (Figure 80C).

From the ostracoderms evolved the **gnathostomes**, the jawed fish, at the beginning of the Devonian. The construction of jaws and teeth from the bones of the **gill arches** was a major evolutionary step, and it allowed fish to become the first major predatory animals. The gills in fish were held open by paired bones in the gill arches: the furthermost anterior pair became teeth, the third pair became the upper and lower jaws, and the fourth pair thickened to become part of the supporting skull. Research in the 1990s on the origins

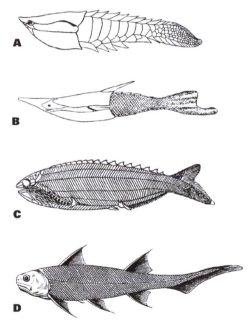

Figure 80 Early fishes. A. *Anglaspis heintzi*; B. *Errivaspis rostrata*, two early heavily anteriorally armoured Devonian heterostrachian jawless fishes; C. *Rhyncholepis parvulus*, an early poorly armoured anaspid jawless fish from the Silurian of Norway; D. *Diplacanthus*, an acanthodian from the Devonian of northeast Scotland.

A–C after Moy-Thomas, J.A. 1939. *Palaeozoic Fishes*. Methuen, London.

and function of jaws sheds some doubts on this scheme; some authors suggest that jaws increased the flow of oxygen rather than performing as they do today.

The most primitive jawed gnathosomes were the **placoderm** fishes, which appeared in the late Silurian but which lasted only until the end of the Devonian. They were very heavily armoured around the head, and possessed paired pectoral and pelvic fins. The largest of these fierce animals was *Dunkleosteus*, which reached over ten metres in length (Figure 81). Spectacular three-dimensional species of this genus and over twenty others have been prepared from the Gogo Limestone of Western Australia. Most of the placoderms were probably bottom-dwelling predators.

The **acanthodians** or spiny-skinned fish were another group of gnathostomes that evolved in the Ordovician and had a long fossil record, albeit with a dwindling diversity into the Permian. They shared features with later sharks in having a cartilaginous skeleton, but differed in having a skin in which small bony

Figure 81 A reconstruction of the head of *Dunkleosteus*, a huge Late Devonian placoderm (FOSP).

plates were arranged. In the Silurian they inhabited marine environments, but by the Devonian various taxa such as *Diplacanthus* from northern Scotland had become exclusively freshwater (Figure 80D).

The sharks or **chondrichthyans** evolved in the Ordovician, but like many members of this group, are poorly preserved on account of the skeleton being made of **cartilage**, which by and large does not fossilise. The skeleton of some later sharks had calcified cartilage, but the fossil record of most is confined to teeth, spines and rare scales. These animals diversified in the Mississippian, and much information about this time and their ecology is derived from collections from Bear Gulch, Montana and from central Scotland. Many of the sharks developed impressive spines on their shoulders, and an examination of their button-shaped teeth has suggested that many were **durophagous**, that is, they were capable of eating organisms with a hard exoskeleton or shell. Some of the most impressive and frightening fossil teeth in geological history are those of *Carcharocles*, a 20 m monster that was related to the Great White Shark (Figure 82). These teeth have been located in Pliocene to Miocene deposits, especially from Malta and southeast USA, and from these the size of the extinct shark has been extrapolated.

The bony fishes or osteichthyans include two major groups: the **actinopterygians** (or ray-finned fishes) (Figure 83A) and **sarcopterygians** (or lobe-finned fishes). They adopted heavy internal bone skeletons to provide a base for the attachment of strong muscles. They have a swim bladder that regulates buoyant density, and an operculum that covers and pumps water over the gills. In today's oceans 95% of all fish are bony fishes, and

Figure 82 Tooth with serrated edges of *Carcharocles*, a massive shark from the Miocene of North Carolina, USA.

the majority of these are ray-finned in which the fins are given added strength through the insertion of thin spines. Lobe-finned fishes are rare in modern oceans. In 1938 an example of one was recovered from East Africa and later named *Latimeria* after the museum curator who first brought it to the attention of a noted **ichthyologist** (Figure 83C). Detailed study of its anatomy revealed that it was a coelacanth, thought to have been extinct for over 100 million years. This caused widespread astonishment in the scientific world, which hailed the fish as a '**living fossil**'. It can also be regarded as an example of a **Lazarus taxon**.

Lobe-finned fishes such as *Eusthenopteron* from the middle to upper Devonian are regarded as the ancestors of the tetrapods, which developed limbs capable of supporting the body while out of water. An early stage in

Figure 83 A. *Vinctiferi comptoni*, a teleost fish from the Santana Formation, Cretaceous of Brazil; B. *Ephippus rhombus*, a ray-finned bony fish from the Eocene of Bolca, northeast Italy; C. *Latimeria chalumnae*, a coelacanth netted off East Africa in 1938. This group of lobe-finned fish was long thought to have been extinct.

this development was the utilisation of paddle-like fins to thrash around in shallow-water ephemeral ponds and lakes, as illustrated by the morphology of the Gogo fish *Gogoanasus*. The transition between fish and the tetrapods and the ability to move about on land remains one of the most important innovations in the fossil record, and opened up a large number of new environmental niches to exploitation.

Some of the oldest exceptional fish faunas are those preserved in the Devonian of Orkney, Shetland and Caithness in northern Scotland, whose diversity was first brought to wider attention by the stonemason and author Hugh Miller. These fish lived in a large freshwater lake that experienced changes in water levels at different times. This would have altered salinity and food availability, alterations that killed most of the fish stocks alive at the time. Sedimentation then covered the carcasses, which ultimately fossilised. The Gogo Formation of Western Australia and the Santana and Crato Formations of Brazil are interesting, as three-dimensional fish are preserved in rocks of late Devonian and Lower Cretaceous age respectively (Figure 83A). Many fish have been preserved in the finely bedded lithographic limestones of Solnhofen (Jurassic) (Figure 37C, p. 49), the Haqel district in Lebanon (Cretaceous), the Green River Formation of Wyoming, USA (Eocene), and from Bolca near Verona in northeast Italy (Middle Eocene) (Figure 83B). Fine detail such as the outline of soft tissue and individual muscle fibres is preserved. While many of the beds at Green River and at Bolca are barren, numerous fish are found preserved in beds at certain levels. These fossiliferous bands represent times of **mass-mortality** in the fish population, which may have been caused by evaporation or the development of algal blooms that would have starved the lakes.

Fossil fish have long captured the imagination of collectors. The Bolca fish were first discovered in 1517 during reconstruction of the citadel at Verona, for which slabs from Bolca were used for paving. These slabs contained impressions which were soon recognised as fossilised fish. The significance of the discovery was soon appreciated, and specimens have found their way into many museums and collections worldwide. The Solnhofen fish have been known since the 1600s. Among the major collectors of the nineteenth century were two young aristocrats who spent their time collecting fish fossils, having been encouraged to do so by Louis Agassiz (1807–1873) whom they met at Neuchâtel, Switzerland while making their Grand Tour. William Willoughby Cole (1807–1886), the third Earl of Enniskillen, and Sir Philip de Malpas Grey Egerton (1806–1881), the 10th Baronet, assembled a large collection with each man retaining one half (part or counterpart) of the split specimen. During their lifetime Cole's collection was housed at Florence Court in Ireland and Egerton's at Oulton Park, Cheshire in England. Later both collections arrived at the Natural History Museum in London, where the individual parts of the fish specimens were reunited. Between 1833 and 1843 Agassiz published *Recherches sur les poissons fossils*, a major monograph in five parts on fossil fish, which drew on the major collections held in public and private museums at the time.

2.13 Tetrapods and Amphibians

Tetrapods were the first animals capable of living on land, and it is thought that they were derived from the lobe-finned fishes some time during the Devonian. They began to develop strengthened ribcages which would later facilitate thoracic breathing, and a number of them, including the Devonian form *Ichthyostega*, even had eight toes on their stout limbs. One important feature of **tetrapod** limbs was that the forelimb could bend backwards at the elbow while the hind limb could bend forwards at the knee. This ultimately facilitated walking.

Some evidence about the mode of life of these animals is known from **trackways** found in Ireland (Figure 84) and Canada. The Irish animal was about a metre in length and had slightly larger hind limbs (**pes**) than forelimbs (**manus**) and drag marks between the **tracks** that it made suggest that its fish-like large tail dragged behind it. Like all early tetrapods it would have found walking on land difficult, but movement through water helped support the body mass.

Amphibians (Class **Amphibia**) have a sparse fossil record that extends back to the early Mississippian. Today they number frogs, newts and salamanders among their ranks, and all have to live in or at least close to water in order to breed. Early fossil amphibians had well-developed legs, although some groups paradoxically lost their limbs during the Pennsylvanian. They had lungs, and pulmonary veins pumped oxygenated blood to the partially divided heart. Amphibians reached their acme during the Pennsylvanian, at which time they were ideally suited for life in the coal swamps that were widespread throughout the equatorial and tropical regions of the globe. Some of the finest assemblages of amphibians date from this period and have been recovered from coal sequences in Canada, the Czech Republic and Ireland (Figure 85A). Others, including the large and impressive *Eryops megacephalus*, have been found in Permian

Figure 84 Trackway made by a Middle Devonian tetrapod. Valentia Island, Co. Kerry, Ireland.

Figure 85 A. *Keraterpeton galvani*, a small amphibian from the Pennsylvanian of Jarrow Colliery, Kilkenny, Ireland. This animal was first described by Thomas Henry Huxley, the noted supporter of Charles Darwin's theory of evolution; B. *Eryops megacephalus*, a large 2 m long amphibian from the Lower Permian of Texas, USA (AMNH).

successions (Figure 85B). Frogs are known since the Triassic, and have a strengthened hip and elongated hindlimbs adapted specially for jumping.

Some amphibian groups developed a fully terrestrial lifestyle, and by the Triassic newer, more advanced stocks had evolved and a number of these, the reptiliomorphs such as Seymouria, showed some characteristics of later reptiles.

2.14 Reptiles

Reptiles contain a wide diversity of animals including the extinct dinosaurs, pterosaurs, as well as three still extant groups: the turtles, the lizards and snakes, and the crocodiles and alligators. They are all characterised through the possession of a number of shared features that include dry, scaly skin and **thoracic breathing**. However, the most important evolutionary development that made the reptiles successful terrestrial organisms was the origination of the **amniotic egg**, which was waterproof. Unlike amphibian eggs, the amnion layer protects the embryo and its foodsource, the yoke, from drying out. This allowed these animals to breed on land for the first time in geological history, and importantly, the young, when ready to hatch, were well developed. Although some reptiles later evolved from their terrestrial ancestors to a marine lifestyle, those, such as the turtles and crocodiles, that spent most of their time in water, were not dependent on it in order to breed. Another feature of many, but not all, reptiles was the arrangement of the limbs, which were progressively placed beneath the main body mass, which was thus effectively supported. For the dinosaurs this was a major factor in some groups attaining enormous size. In time, the muscular arrangements of the reptiles became modified relative to that of amphibians, so that they were capable of exerting a large bite pressure. This, too, aided the predatory lifestyle that many of the dinosaurs were to follow.

The earliest reptile is *Hylonomus lyelli*, which was discovered by Sir William Dawson and Sir Charles Lyell in 1852 in middle Pennsylvanian rocks. They found that the unfortunate animal had become trapped within a hollow stem of the lycopod tree *Lepidodendron*, which subsequently became infilled by sediment. This was the first true **amniote**. Several other early tetrapods began to acquire reptilian traits, including *Westlothiana lizziae*, which was discovered near Glasgow in Scotland.

Reptiles increased in importance during the Permian and diversified rapidly in the Triassic at a time when **Pangaea**, a supercontinent that comprised most of the continental plates, was experiencing warm and dry climatic conditions.

The major groups of reptiles can be distinguished through an examination of their skulls (Figure 86). The **anaspids** had a skull with a large eye socket but no temporal opening behind it (Figure 86A). This is the most primitive skull state and is seen in the earliest reptiles and in turtles and tortoises today. The **synapsids** have a skull with a single large temporal opening behind the eye socket (Figure 86B) and are represented by the mammal-like reptiles (*see* Section 2.15). In the **diapsids** the skull has two temporal openings behind the eye socket (Figure 86C) and these include the dinosaurs, pterosaurs, lizards, crocodiles and birds. The last reptilian skull type is that of the **euryapsids**, which have a single small opening

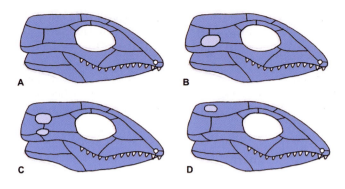

A

B

C

D

Figure 86 Reptilian skulls. A. Anaspid skull; B. Synapsid skull with a single temporal opening; C. Diapsid skull with two temporal openings; D. Euryapsid skull with a small opening high up on the skull.

After Swinton, W.E. 1973. *Fossil Amphibians and Reptiles.* British Museum (Natural History), London.

high up on the skull behind the eye socket (Figure 86D). The swimming reptiles such as ichthyosaurs and plesiosaurs are euryapsids.

The anaspid turtles (Figure 87) and tortoises are members of the Order Testudines, which are easily recognised on account of their large dome-shaped carapace made of a number of fused bony plates. They have a geological history that began in the Triassic when they lived on land. Later in the Mesozoic some took to the oceans, where Cretaceous turtles thrived, reaching over 2.5 m in length. These long-lived animals had teeth in at least one Triassic form, but by the Paleogene and in modern examples these are lost.

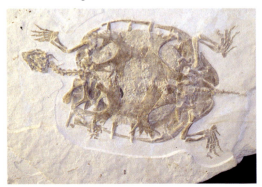

Figure 87 *Eurysternum wagleri*, a turtle from the Jurassic of Solnhofen, Germany.

In many museums specimens of Jurassic swimming reptiles such as the ichthyosaurs, plesiosaurs and pliosaurs are displayed in prominent positions, and not without good reason. These euryapsids are beautiful animals that have captivated people since the finds made by Mary Anning around Lyme Regis in Dorset, England, in the early 1800s. The ichthyosaurs superficially resemble modern-day dolphins and porpoises in that they possess a streamlined body with a long snout carrying sharp peg-like teeth and four paddle-like flippers, but anatomically they are quite different. The plesiosaurs had small skulls and very long necks, while the pliosaurs had large skulls and shorter necks. The latter are best exemplified by the Jurassic animal *Rhomaleo-saurus cramptoni* that was excavated from an alum quarry at Kettleness in Yorkshire in 1848, and which reached 7 m in length (Figure 88). These animals propelled themselves gracefully through the water using their flippers. A great deal is known of the diet of these marine reptiles through a study of their coprolites; dissection of these fossil faeces show that both fish and cephalopods were favoured prey. At least some of the marine reptiles could bear their young live, which was proved on the

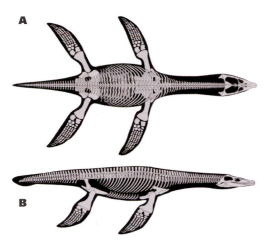

Figure 88 *Rhomaleosaurus cramptoni*, a large pliosaur from the Jurassic of Whitby, Yorkshire. Top: ventral view; bottom: side view (© Adam Stuart Smith).

discovery of a specimen of a *Stenopterygius* female from Holtzmaden in Germany that died in the throes of giving birth.

The most diverse group of reptiles were the diapsids, which rapidly radiated in the Triassic, and exploited most terrestrial niches available to them, including the air. The dinosaurs are the best known of this group and have been divided into two major **clades**, the **Saurischian** ('lizard-hipped') and the **Ornithischian** ('bird-hipped') dinosaurs. In the former group, which includes the giant herbivorous **sauropods** such as *Barosaurus* (Figure 89A), *Brachiosaurus* and *Diplodocus* and the predatory **theropods** such as *Tyrannosaurus* (Figure 89B), the two bones of the pelvis, the pubis and the ischium point in opposite directions. In the latter group, which is considered to be more advanced, both bones point backwards, and all of these animals in the group are herbivores and include the horned dinosaurs *Triceratops* (Figure 90A) and the plated dinosaur *Stegosaurus* (Figure 90B).

Figure 89 (left and below) A. Saurischian dinosaurs: *Barosaurus*, a large sauropod. This animal is mounted in a raised defensive position (AMNH); B. *Tyrannosaurus rex* (AMNH).

Figure 90 Ornithischian dinosaurs: A. *Triceratops horridus*; B. *Stegosaurus* (AMNH).

Dinosaurs are probably the most studied fossil group. In the US disputes in the late 1800s between early collectors such as Othniel Marsh (1831–1899) and Edward Drinkwater Cope (1840–1897) led to what became called the 'Bone Wars' during which each protagonist attempted to retrieve bones for their own exclusive study. More recently, dinosaur finds have resulted in court cases where the ownership of the specimens has been disputed. Victorian collectors spent large sums of money excavating specimens and returning them to the large provincial museums in Europe and North America. Many fine sauropod specimens were retrieved by German and British scientists from Tanzania, while in southern Alberta, Canada, the abundance of fossils led to the creation of Dinosaur Provincial Park. Andrew Carnegie, who had made his money in the steel mills of Pittsburg, even had copies of *Diplodocus* made which he sent to museums in London and Madrid and elsewhere. In Belgium over thirty specimens of *Iguanodon* were discovered fossilised in a coal mine, and excavation of these allowed scientists to re-evaluate the stance of this animal, demonstrating that this was **bipedal** and not **quadrupedal** as previously thought. Over the last two decades many new exciting dinosaurs have been unearthed in remote parts of China and South America, and today a week barely passes without some notice in the popular press announcing a new dinosaur find.

All of these finds have provided vertebrate palaeontologists with a large data set, and in the last thirty years great efforts have been made to understand more fully the actual anatomy and morphology of the animals, their habits and their ecological meaning. Scientists are not content simply to name the fossils, but they want to bring them to life, so to speak, so that the bones also reveal something of the times in which the fossil organisms lived.

Trackways made by dinosaurs are known from many localities worldwide. The first were discovered in Connecticut in the 1830s and were produced by several small bipedal dinosaurs, but interpreted at the time as having been made by a large prehistoric bird affectionately referred to as 'Noah's raven'. Other trackways from North America, South America, Europe and elsewhere have revealed that many dinosaurs lived in social groups, and palaeontologists have even been able to work out the walking speeds, patterns, and weight of the animals that made them.

Figure 91 Two *Protoceratops* skeletons, small dinosaurs from the Cretaceous of Mongolia together with a clutch of their eggs (AMNH).

Figure 92 *Pterodactylus spectabilis*, a pterosaur from the Jurassic of Solnhofen, Germany.

The recovery of dinosaurs' eggs laid by *Protoceratops* and *Oviraptor* at Flaming Cliffs in Mongolia in the 1920s and later that century (Figure 91), and then more recently in France and China, have suggested that some dinosaurs, including *Oviraptor*, may have had a strong parental instinct. **Dinosaurs**, like many other groups, were fatally wiped out at the end of the Cretaceous.

Their limb morphology allowed the pterosaurs (Figure 92) to take to the air. The finger bones of the forelimb became greatly elongated and held a thin membrane between it and the body. This was used in a flapping motion to effect rudimentary flight. *Quetzalcoatlus* and *Hatzegopteryx*, two Cretaceous pterosaurs, had wingspans of approximately 11 m while that of *Nemicolopterus*, another pterosaur of much the same age from China, extended to only 25 cm. Other diapsids evolved into the modern-day birds (*see* Section 2.14).

The crocodiles, which date back 200 Ma, adopted a partially aquatic lifestyle. Mature,

Figure 93 Skull and long snout of the Lower Jurassic teleosaur crocodile *Steneosaurus bollenses* from the Holtzmaden Shale of Boll, near Stuttgart, Germany.

fully grown examples of the Jurassic teleosaur crocodile *Steneosaurus* (Figure 93) reached 3 m in length, and like one half of the crocodilians, had prominent **longirostrine** (long and thin) jaws with eyes held on top of the skull and limbs held out from the lateral sides of the body. The other half had shorter **brevirostrine** jaws. In Britain, specimens of *Steneosaurus* and *Metriorhynchus* were excavated from the Oxford Clay (Jurassic) of Peterborough in the 1990s, and these are now held by Leicestershire Museums. Earlier crocodilian finds in Britain included Jurassic eggs from the Cotswolds.

The synapsids included *Dimetrodon*, which trapped heat by means of a large sail that extended from its backbone. Within this sail blood vessels would have been close to the surface, and so the animal heated up sufficiently fast to the temperature at which it could move reasonably rapidly. This adaptation may have led to **endothermy**, being warm-blooded, which was a major advantage utilised by the mammals, which evolved from a synapsid reptile stock during the Triassic.

Although modern-day reptiles may appear to be rather insignificant when viewed alongside the huge diversity of mammals with which they share the planet, it is worthwhile remembering that they had a very long tenure, during which time they were the pre-eminent organisms on Earth.

2.15 Birds

It is probably true to say that most people, if they think about it, are quite familiar with the anatomy of birds. They will have either carved numerous chickens or seen them being carved. Modern birds have asymmetrical feathers, thin and hollow bones, a large sternum or breastbone that supports the strong muscles needed to power flight, and a short tail, or no tail at all.

The oldest bird is *Archaeopteryx lithographica* (Figure 37B, p. 49) from the Solnhofen Limestone of Bavaria in Germany, and is Jurassic in age (150 Ma). Since 1860 eight specimens and one feather have been discovered. This bird had solid bones, claws on its forelimbs, teeth rather than a beak, and symmetrical feathers. It has been interpreted as being capable of flight, albeit rather clumsy flight. In the 1970s the American scientist John Ostrom established ancestry of the birds with small carnivorous dinosaurs, and since then new finds have corroborated his ideas.

In the early Cretaceous rocks of Liaoning Province in northeast China, a number of so-called '**dinobirds**' have recently come to light, and these confirm the link with dinosaurs such as *Deinonychus*. One of these birds, *Sinosauropteryx*, which dates from 125 Ma, was covered in fluff that seems to have been a precursor of feathers, and may have been used for insulation (Figure 94A). Other 'dinobirds'

Figure 94A *Sinosauropteryx,* a small 'fluffy' dinobird from the Cretaceous of Liaoning Province in northeast China.

Figure 94B The skull of a bird complete with coloured feathers from the Fur Formation of the Eocene of Denmark.

such as *Confuciusornis* had well-developed feathers, a beak and a shortened tail, and as such is one of the first modern birds. Additional early Cretaceous birds have been found in the Las Hoyas Limestone of eastern Spain.

Modern birds number over 10,000 species, of which the perching birds make up just over half the number. Some experts have argued that they originated in the Cretaceous, but many records from the Mesozoic are problematic, although they suggest that some primitive modern bird groups did develop at that time. Other experts, drawing on research on fossils found in the Eocene London Clay and sequences of the same age in Denmark (Figure 94B), have countered, saying that most, and the more derived modern birds, originated after the Cretaceous–Paleogene boundary. Some of the most exciting recent studies on fossil birds have led to the discovery of pigmentation in feathers from the Cretaceous and the Eocene (Figure 94B).

2.16 Mammals

The **Cenozoic** has been called the 'age of the mammals' due to their dominance after the mass extinction event at the end of the Cretaceous. Today there are some 4000 different mammalian species.

The first mammals evolved about 220 Ma, at the same time as dinosaurs, and these animals shared three characteristics with modern mammals. They possessed mammary glands, hair, and had bones in the middle ear that were derived from jaw bones. Other characteristics of mammals are that they are endothermic, which allowed them to regulate the temperature of their blood. This made them far more active than the reptiles and also allowed them to inhabit colder environments, an adaptation that would prove particularly important later, during prolonged glaciations from the Oligocene onwards.

During the Permian, a group known as the **theraspids** appeared and managed to survive the end-Permian extinction to give rise to the **cynodonts** during the Triassic. These **mammal-like reptiles** had differentiated teeth, a synapsid skull (Figure 86B) and a **hard palate**. The latter was a useful adaptation, as it allowed them to eat and breathe at the same time; something most reptiles cannot do. Some of the finest cynodont examples, which date from about 230 Ma, have been located in the Karoo region of South Africa. In their evolutionary journey, successive cynodonts gained ever more mammalian features, such as a shortened tail, smaller eye sockets, and a lower jaw almost exclusively made up of the **dentary bone**. Cynodonts continued until the end of the Jurassic.

The first true mammals appeared at the very end of the Triassic and into the early Jurassic. These were small, whiskered, insectivorous, nocturnal creatures such as *Megazostrodon* (Figure 95) known from South Africa and *Morganucodon*, whose bones have been found in sediment preserved in fissures in rocks in

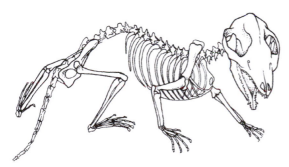

Figure 95 *Megazostrodon*, an early mammal from the Lower Jurassic of South Africa. Left: skeleton; bottom: restoration.
After Benton, M.J. 2005. *Vertebrate Palaeontology.* Blackwell Publishing, Oxford.

southwest England and in Wales. At the same time a group appeared that ultimately gave rise to the **monotremes**, a primitive egg-laying group of mammals exemplified by the modern-day Duck-billed Platypus and the Echidnas.

At some point in the Cretaceous, around 100 Ma ago, two major groups of mammals evolved, each of which had different reproductive styles. These were the **marsupials** (Suborder **Metatheria**) whose descendants are known from modern Australia, and the **placental mammals** (Suborder **Eutheria**). The metatherians gave birth to physically immature young, which then developed further within the mother's pouch. They were successful in the past and radiated throughout **Gondwana** before it began to split into what is now South America, Africa and Australia. Today the mammalian faunas are dominated numerically by the eutherians. The young of placental mammals develop to a relatively advanced state inside the mother, with nutrients supplied not by an egg, but by a network of blood vessels and membranes called the placenta.

Following the disappearance of the dinosaurs and other reptiles, the mammals were free to exploit the vacant terrestrial and marine niches, and a rapid diversification followed, so that by the Eocene more than twenty major groups had appeared. These included the Rodentia, the Lagomorpha (rabbits and hares), the Perissodactyla (Rhinos such as *Indricotherium*, which is the largest mammal ever known, and Horses), the Artiodactyla (Hippos, Camels (Figure 96), and Deer (Figure 97)), the Carnivora (Cats, Dogs, Bears, and Wolves), the Chiroptera (Bats), the Proboscidea (Elephants (Figure 98)), the Cetacea (Whales such as *Basilosaurus*, which measured 25 m long, and Dolphins) and the Primates (Monkeys).

Figure 96 The dwarf camel *Stenomylus* that stood only 60 cm high and which was native to North America (AMNH).

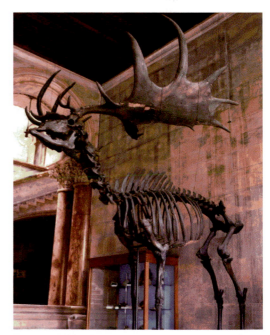

Figure 97 *Megaloceros giganteus*, the Giant Irish Deer from the Pleistocene of Lough Gur, Co. Limerick, Ireland. The antlers on this male spanned 3.5 m.

Figure 98 The mammoth *Mammuthus jeffersoni*, a non-woolly mammoth, from the Pleistocene of Jonesboro, Grant County, Indiana, USA (left), and the Pleistocene mastodon *Mammut americanum* found in 1845 in a peat bog near Newburgh, New York, USA (right) (AMNH).

The most primitive placental mammals were some South American taxa such as the armoured *Glyptodon* and *Megatherium* (the giant ground sloth) which are well represented in museum collections. The more advanced placentals can be conveniently divided into the **carnivorous** groups and the **herbivorous** ungulates, many of which attained enormous size.

Three million years ago or so, the link between the north and south American continents was established along the Panama Isthmus, and quickly there was migration of mammals between the two landmasses. Mammoths, mastodons, dogs, wolves, deer, camels and others all moved south and passed sloths and other southern animals moving north. While it has been argued by some authors that competition for resources led to the demise of most southern marsupials that were less able than their northern counterparts, Michael Benton has written suggesting that in fact most southern groups were already in decline.

Parallel-evolution has been documented in a number of groups, with two sabre-toothed cats *Smilodon*, a placental and *Thylacosmilus*, a marsupial, broadly resembling each other. No doubt they were major predators in their respective environments.

If the mammals were so successful during the Paleogene, why has their diversity declined, and why are modern taxa significantly smaller than their earlier counterparts? Many certainly succumbed to hunting by early Man, and in the Neogene they would have been adversely affected by the glaciations in the Pleistocene, when food in high latitudes would have been at a premium. Soon, gone were the herbivorous giants and the nimble carnivores, and today we are left with a very reduced mammalian fauna with few large animals other than rhinos and elephants that faintly reflect and suggest the dominance of their past ancestors.

2.17 Hominids and Hominins

The Family Hominidae, to which group modern man, as well as the other great apes such as gorillas, chimpanzees and orangutans belong, have a long history reaching back some 50 Ma. Taxonomically the **hominids** are part of the larger group, the **Primates**, which include the apes, chimps, monkeys, which have grasping hands, nails (not claws) and **stereoscopic vision**. The earliest Primates were the **Prosimians**, which appeared 70 Ma ago and today include the rarely seen Aye-aye from Madagascar. These were small, mainly **nocturnal** animals that lived off berries and insects. Twenty million years later both the Lemurs and the New World Monkeys appeared. It has been shown that basal primate *Darwinius massillae* (nicknamed 'Ida') from the Grube Messel of Germany has lemur-like features, but its position in the phylogeny of the primates as a link between the prosimians and the more derived groups remains a matter of debate. New World Monkeys possessed a prehensile tail, and open, wide nostrils (Spider Monkeys and Golden Tamarins). The geological history of the Old World Monkeys begins 40 Ma ago, and these differ in having downward-pointing nostrils and tails (Baboons and Colobus monkeys). The first apes date back to 50 Ma, and the first organisms that closely resemble modern man to the Miocene some 22 Ma ago.

What sets Primates apart from other groups? They developed a number of methods of moving from place to place. They could leap from tree to tree (Lemurs); others mastered **brachiation**, using their arms to swing through forests; quadrupedal walking; knuckle-walking (e.g. Chimpanzees); and finally fully upright walking as in the hominids.

Modern humans are closely related to the Chimpanzees, with whom they share the Tribe Hominini. The Subtribe Hominina now contains one living species – modern man *Homo sapiens sapiens*, but also a number of now extinct antecedents. Within the phylogeny or geological history of the hominins there were a number of morphological adaptations that allowed the group to become highly successful and ultimately to spread out throughout the globe (Figure 100). Probably the prime adaptation related to diet, which saw **hominins** change from a fruit-dominated diet to that of an omnivore that could feed on almost any foodstuff, and this coincided with climatic changes in the late Miocene that resulted in lush forests of east Africa being replaced with open grasslands. This change in feeding habits was facilitated by a change in the shape of the jaw, where early hominins and apes had a rectangular-shaped jaw while in later hominins this had become more parabolic. Equally important were changes to the shape of teeth – in modern man canines are reduced and molars are larger. Over time the brain capacity of the apes and hominins increased from 400 cm^3 in chimpanzees to 1400 cm^3 in modern man. The dexterity of hominin hands and their ability to utilise materials available to them as tools also set them apart from their

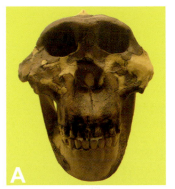 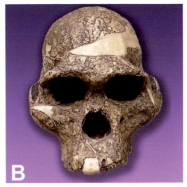 

Figure 99 Hominin skulls. A. *Paranthropus boisei*, Olduvai Gorge, Tanzania (replica LNHM); B. *Australopithecus africanus*, Sterkfontein Cave, South Africa (replica TMP); C. *Homo habilis* discovered at Koobi Fora, Kenya (replica).

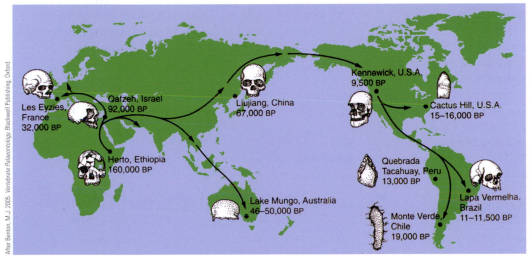

After Benton, M.J. 2005 *Vertebrate Palaeontology* Blackwell Publishing, Oxford

Figure 100 Radiation of *Homo sapiens* out of Africa and across the globe. BP = before the present.

ancestors. The first evidence of tool-making goes back to *Praeanthropus (Australopithecus) afarensis* some 3.4 Ma ago.

The evidence for the phylogeny and relationships of early hominins is sparse, as few fossils have been found. Nevertheless several important finds have been made, and these are documented below. In recent years advances in genetics have resulted in the recognition of new hominin species from small fragments of bone, and no doubt in a short time our understanding of our ancestry will become more fully known.

Synopsis of hominin history

Proconsul – 22 Ma. Fossils first found in 1909 in Miocene deposits of Kenya. On the evidence of its teeth it had a herbivorous diet. This is just one of a number of possibilities that may be the root-stock of the Hominins, although this is disputed by some experts.

Sahelanthropus tchadensis – 7 Ma. Found in 2001. Cranial fragments found in Chad show it to have had a brain capacity of 350 cc. Possibly on the ancestral line of modern man.

Orrorin tugenensis – 5.6 to 6.2 Ma. Discovered in the Tugen Hills of Kenya. Bipedal in stance, 90 cm high. A forest dweller, eating fruit, nuts, and meat.

Ardipithecus ramidus – 4.4 Ma. Nicknamed 'Ardi'; found in the Afar Rift in northeast Ethiopia in the 1980s and first reported in 1994, and again in 2009. Stubby teeth; dwelling in cool, humid forests and eating fruit, nuts, mammals and birds. 120 cm high, weighing 50 kg.

Australopithecus africanus – 3 Ma. Found in South Africa by Raymond Dart in 1924. Large skull, small thighs, upright stance. Became extinct 1 Ma ago (Figure 99B).

Praeanthropus (Australopithecus) afarensis – 3.4 Ma. 'Lucy', found in Ethiopia. Broad eyebrow; 100 cm high; slight frame. Forest dweller and fruit eater. Earliest user of tools yet known.

Paranthropus (Australopithecus) boisei – 3 to 4 Ma. Found in 1959 by Mary Leakey at Olduvai Gorge, Tanzania. Footprints discovered in volcanic ash dated at 3.5 Ma. Grassland plains dweller (Figure 99A).

Homo habilis – 2.4 Ma. First member of the genus *Homo*. Toolmaker nicknamed 'Handyman'. Found in 1961 at Olduvai. Brain capacity 650–700 cc. 1.3 m high (Figure 99C).

Homo erectus – 1.8 Ma to 750,000 BP (Before Present). These hominins migrated out of Africa, and their fossils have been found in Africa, China (Peking Man), Indonesia (Java Man), and Europe. Brain capacity 900 cc. They discovered fire, and it is thought, developed close social groups.

Homo sapiens neandertalensis – 500,000 BP. Had a brain capacity of 1500 cc (larger than modern man). Rugged features; large eyebrow. Migrated back into Europe after the Ice Age. They were displaced and eventually became extinct after competition with sleeker *Homo sapiens sapiens*.

Homo sapiens sapiens – 500,000 BP to the present day. Modern man. Brain 1400 cc. Mass migration to all of the globe, reaching Europe 32,000 BP, China 67,000 BP, Australia 50,000 BP, and South America 11,000 BP (Figure 100).

2.18 Trace fossils

Ichnology is the study of the behaviour of extinct organisms by examination of their traces, tracks, trails, burrows, or borings. These are left in sediments or in hard substrate of some sort, such as rock, wood, or even the skeletons of other animals, and may be subsequently preserved as trace fossils. They were produced by the activity of animals (usually invertebrates but sometimes vertebrates). Many animals such as worms, containing soft tissues and no hard skeletal tissue, are known only from trace fossils, as after death the animal decays and the body is not preserved. There are many examples of bioturbation in sediments where these soft-bodied animals have burrowed through sediment, disrupting layering and bedding. In some cases even the makers of large traces remain elusive. *Chirotherium*, which means 'hand beast', is known from trackways that were first discovered in Triassic sandstones in Germany (Figure 101), and later in Cheshire in England, Northern Ireland and Africa. All that palaeontologists know about the animal has been determined from its trackways, which have indicated that it was a reptile some 2–3 m in length with a large hindlimb (pes) and a smaller forelimb (manus). Of course reptilian body fossils are known, but as yet none have been shown to be of the animal responsible for *Chirotherium*, although the suspect is considered to be an archosaur, which was a forerunner of the dinosaurs.

Figure 101 *Chirotherium* trackway from the Triassic of Hildburghausen, Germany.

Burrows are excavations made into an unconsolidated substrate (Figure 102A). **Borings** are excavations made into a consolidated substrate, which could include rock or wood, or indeed the skeleton or shell of another organism (Figure 102B). Some bivalves and sponges are **endolithic** (bore into stone) and many may be responsible for causing mechanical breakdown or **bioerosion** of limestone. Before the advent of metal ships, one constant source of worry for sailors was the damage that could be caused by the ship worm (actually a bivalve) *Teredo* which produced borings called *Teredolites* in ship timbers. **Tracks** are imprints on a sediment surface by an animal with legs (Figure 101A). **Trails** are imprints made in sediment by a legless animal dragging its body across a surface. Of course not all traces may be preserved; most tracks are formed in unconsolidated sediments that may not become preserved as part of the fossil record.

Trace fossils are given latinised generic and species names in the same way that body fossils are. For example, in *Oldhamia radiata* (Figure 12, p. 19), *Oldhamia* = genus or **ichnogenus** and *radiata* = species or **ichnospecies.** Regardless of its biological origin, the name often represents the geometry of the trace. Typically, names carry the suffixes *-ichnus*, *-craterion*, *- ichnites* [footmarks], *-opus*, but not all do. The use of the suffix *-ichnus* to distinguish trace fossils from body fossils was first suggested in 1853.

Trace fossils may be classified according to their physical characteristics, or more usefully according to the behavioural patterns of the animal that produced them. There are seven such behavioural groups: Repichnia = Crawling traces, Cubichnia = Resting traces;

Figure 102 A. The burrow *Taenidium* (which was once called *Beaconites*) from the Devonian of Largysillagh, Co. Mayo, Ireland; B. Cross-section through various endolithic borings including the wide *Gastrochaenolites* (some with boring bivalves in place) and the narrower *Trypanites* in a Jurassic hardground or rockground of the Mendip Hills, England.

Pascichnia = Grazing traces (Figure 103); Fodinichnia = Feeding traces; Domichnia = Dwelling traces; Fugichnia = escape traces; Agrichnia = farming traces.

Trace fossils have a number of useful applications for geologists and palaeontologists. They can inform about the biological nature of the tracemaker, can suggest information

Figure 103 *Helminthoidea*, a gastropod grazing trace from the Cretaceous of Barcelonetta in the French Alps.

about the environmental conditions under which they were made, and can be useful in biostratigraphy and correlation.

Many traces can be attributed to the organism that made them, but many cannot. Equally it is possible that similar traces have been made independently by different organisms at quite different times in the fossil record. It is also possible that different ichnogenera may be made by the one tracemaker, the characteristics of each being determined by the nature of the substrate. Some polychaete worms are known from the Ordovician of Estonia, and these produced long tubular borings (*Trypanites*) in massive dome-shaped bryozoan colonies, whereas in smaller cylindrical colonies the boring (*Sanctum*) was morphologically different. Vertebrate tracks appeared following the emergence of the first tetrapod vertebrates (amphibians) in the Devonian Period. Good examples are known from Canada, western Ireland, and the oldest have recently been reported from Poland. Tracks of reptilian, avian, and mammalian tracemakers are also common in the geological record. Vertebrate tracks show whether the animal was bipedal (walked on two legs) or quadripedal (walked on four legs). Tracks can yield important information on the size of the trackmaker, on its weight, on its pattern of walking, on whether the animal walked, ran or galloped.

Certain assemblages of trace fossils have been shown to be characteristic of particular water depths and environments, and some of these have been consistent through geological time. For example, deep-water abyssal traces from Cambrian rocks are similar to those produced today. Ichnologists who study such traces have identified nine distinctive **ichnofacies** that are based on where they occur in terms of water depth and sediment type and setting: *Scoyenia* ichnofacies – low diversity assemblage in lacustrine environments that may contain king crab and vertebrate trackways; *Trypanites* ichnofacies – borings in limestone made by algae, annelid worms, clionid sponges (*Entobia*), echinoids and bivalves, particularly *Pholas; Glossifungites* ichnofacies – typically found on hardgrounds; *Skolithos* ichnofacies – found generally in intertidal areas where sediment is often replaced (*Skolithos*; *Thalassinoides*; *Diplocraterion*; *Arenicolites*); *Cruziana* ichnofacies – rich assemblage of crawling and burrowing traces. These are found below normal wave base, but not storm base, and include trilobite walking traces (*Cruziana*), resting traces (*Rusophycus*) and starfish resting traces (*Asteriacites*); *Zoophycos* ichnofacies – range of water depths from shallow to deep water (*Zoophycos*); *Nereites* ichnofacies – generally in deep water, in muds and turbidites (*Helminthoidea*); *Psilonichnus* ichnofacies – found in supratidal marshes (*Psilonichnus*; rootlets; vertebrate tracks); and *Teredolites* ichnofacies – borings in wood, often made by the ship worm *Teredo* (*Teredolites*).

Some trace fossils have proved very useful in providing age determinations of sedimentary rocks. Some Cambrian successions contain few body fossils but good examples of *Cruziana* that were made by trilobites as they walked over the sea floor. Some recent reviews have demonstrated that the diversity of ichnogenera increased through the early Phanerozoic – the Cambrian assemblages were less diverse than those of the Ordovician to Pennsylvanian and contained fewer feeding and grazing traces than in later sediments.

The term 'ichnology' was first introduced in about 1830 by the Rev. William Buckland, Professor of Geology in Oxford. He was also responsible for naming *Megalosaurus*, the first dinosaur formally described from Britain, and was interested in various cave faunas and Lower Jurassic **coprolites** (fossil faeces). He became Dean of Westminster, but unfortunately, late in life, he lost his mind and was institutionalised.

Glossary

Abdomen [105]: the posterior portion of the body in arthropods behind the thorax.

Abyssal [33]: areas of the ocean below 4000 m depth.

Acanthodians [118]: a group of extinct fish that lived between the Ordovician and Permian which shared some characteristics of both the cartilaginous fish and the more advanced bony fishes.

Accession number [14]: the catalogue number given to a museum specimen.

Acme [90]: the period of the greatest abundance of an organism or group of organisms.

Acritarchs [22]: a group of microfossils which range from Precambrian times to the present that show affinities with algae and egg-cases.

Actinopterygiians [118]: a major group within the bony fishes characterised by having fins strengthened by bony rays. Informally called the ray-finned fishes.

Adapical [99]: the top, i.e. opposite surface to the mouth in echinoids.

Ahermatypic [71]: a group of corals that lack symbiotic photosynthetic algae and which do not form reefs, unlike hermatypic forms.

Allochthonous [33]: sediments and other material transported to their place of deposition.

Ambulacra [93]: the five sections of the test (shell) in echinoids that contain pores from which soft tube-feet emerge. In blastoids, the area of attachment of the brachioles and where water is passed into the body.

Amino-acids [22]: the building-blocks that make up proteins.

Ammonoidea [79]: a subclass of the Phylum Cephalopoda with a coiled shell and complex sutures. They were abundant in the Mesozoic.

Amniote [123]: a group of animals including reptiles, birds and mammals that produce an amniotic egg.

Amniotic egg [123]: an egg that is waterproof and so can be laid on land. The foetus is surrounded by several protective layers.

Amphibia [121]: a class of the subphylum Vertebrata which developed limbs and lungs that allowed them to move across the land. They include the newts, salamanders and frogs, and have to lay their eggs in water.

Anaspid [123]: a skull that lacked a temporal opening behind the eye socket. Typical of the first reptiles and modern-day tortoises and turtles.

Angiosperms [26]: flowering plants.

Anoxic [48]: containing no oxygen. Can occur in some deep-water sediments or waters or in marine basins with poor circulation.

Anterior, -ally [77]: the front end of a body or shell.

Anthozoa [68]: a Class of the Phylum Cnidaria that contains the corals and the sea anemones.

Apatite [5]: a calcium phosphatic mineral of which bones and teeth are often composed.

Apex Chert [43]: a siliceous rock found in Western Australia dating from 3500 Ma that contains the earliest examples of life on Earth.

Appendages [103]: limbs used for locomotion.

Aptychi [78]: paired plates in cephalopods that cover the aperture or mouth.

Aragonite [5]: a form of the mineral of calcium carbonate $CaCO_3$ frequently found in the marine shells.

Archaean [22]: an Eon, a major division of time between 4000 Ma and 2500 Ma.

Aristotle's Lantern [93]: the feeding jaw apparatus in echinoids comprising five sharp teeth, plates, and muscles situated on the ventral surface or underside.

Arthropods [101]: a phylum of animals characterised by a segmented body and appendages. They include trilobites, crabs and shrimps.

Ascon [65]: the simplest body form in sponges, consisting of a bag-like structure.

Asphalt [50]: a thick viscous semi-solid hydrocarbon, natural accumulations of which have acted as a preservative of fossil material, as at Rancho La Brea, California.

Asteroidea [98]: a Class of the Phylum Echinodermata containing the starfish and brittlestars.

Atmosphere [22]: the layer of gases surrounding the Earth. This over time has changed composition due to various biological and non-biological factors such as photosynthesis and volcanic eruptions.

Autochthonous [33]: sediments and other material formed in their place of deposition.

Autoecology [33]: the study of the relationship of a single species of plant or animal to the environment in which they live.

Autotheca [110]: in graptolites the chamber that houses the polypide.

Baltic Amber [48]: fossilised resin from pine trees produced 35 million years ago and now preserved in sediments around the Baltic Sea and eastern North Sea.

Banded Ironstone Formation [23]: iron-rich deposits formed 2.5 billion years ago during the Proterozoic.

Bedding plane [10]: the flat horizontal surface of a bed of sedimentary rock.

Beds [12]: distinct horizontal units of sedimentary rock.

Benthic [34]: a term used for animals that live in, at, or just above the seabed.

Binocular microscope [15]: a microscope with two objectives used for viewing specimens.

Binomial [18]: a name for an organism made up of two parts: a genus name followed by a specific name, i.e. *Baculopora megastoma*.

Bioerosion [5]: breakdown of rock and skeletal parts by the actions of living organisms. Many bivalves and sponges are capable of boring into rock, while other groups erode using chemicals.

Biostratigraphy [36]: the part of stratigraphy that utilises fossils to date rock units.

Biostratinomy [5]: the events and time between death of an organism and the burial of its remains.

Biota [45]: the flora and fauna of a particular environmental setting or place.

Biozones [36]: a short time unit identified by a distinctive fossil or group of fossils.

Bipedal [126]: walked on hind limbs.

Biramous [103]: limbs having two distinct portions as in trilobites, whose limbs were divided for walking and for use as gills.

Bitheca [110]: a small cup in dendroid graptolites whose function is unclear; smaller than the adjacent autotheca that housed the polyps.

Bituminous [47]: containing the tar-like hydrocarbon bitumen, as in the marls around Holtzmaden, Germany that led to the preservation of soft tissues of many organisms including large marine reptiles.

Bivalvia [84]: a class of the Phylum Mollusca characterised by having two hinged valves, usually, but not always, symmetrical.

Blastoid [96]: a member of a class of the Phylum Echinodermata that is attached to the substrate by means of a stalk and with a calyx with ambulacra and delicate brachioles.

Body chamber [77]: the final portion of the shell in cephalopods and gastropods where the animal resides. It is separated from earlier chambers by a septum.

Bolide [26]: a meteor that explodes above the Earth's surface.

Boring [138]: a tunnel and excavation made into a hard material such as shell by a small animal.

Brachials [97]: the arms of crinoids.

Brachiation [134]: the process of using arms to move from place to place. Many monkeys move in this way.

Brachidia [89]: coiled skeletal structure in brachiopods that supports the lophophore.

Brachioles [98]: delicate arms in blastoids.

Brachiopoda [89]: members of a phylum of marine animals having two asymmetrical valves often attached to the substrate. Common in the Palaeozoic, they declined during the Mesozoic and today are found confined to deep continental shelf areas.

Brackish environments [63]: areas where the salinity of the water is intermediate between fresh-water and fully marine.

Brevirostrine [128]: having a short snout as in some crocodilian groups.

Brood chambers [74/104]: structures in bryozoans and in some trilobites where larvae are held and matured before being released into the surrounding water.

Bryozoa [72]: colonial sessile organisms with individuals housed in discrete chambers; feed by means of a tentacle crown (lophophore).

Buccal cavity [116]: the mouth cavity in the head containing the teeth and tongue.

Burgess Shale [44]: a Middle Cambrian sedimentary unit in British Columbia, Canada, containing a diverse assemblage of exceptionally preserved soft and hard-bodied organisms.

Burrow [138]: a tunnel and excavation made into unconsolidated sediment by a small animal.

Calcite [5]: a mineral composed of calcium carbonate $CaCO_3$ that makes up limestone and is utilised by many organisms to build their skeleton.

Calcium compensation depth [64]: the depth between 4000 and 5000 metres in the ocean at which calcium carbonate goes into solution. This occurs due to the great pressures at depth.

Calice [69]: the cup-shaped depression on top of a coral skeleton in which the soft-tissued polyp is situated.

Calyx [96]: the cup on top of the stem made up of the theca and the tegmen in a crinoid or blastoid in which the body is contained.

Cambrian [22]: a division of geological time in the Palaeozoic Era between 542 and 488 million years ago.

Cameral deposits [78]: calcitic deposits in the lower sides of chambers in orthoconic nautiloids that keeps them correctly orientated in the sea water.

Carapace [34]: the hard outer skeleton of an arthropod.

Carboniferous [40]: a term now encompassing the two geological periods Mississippian and Pennsylvanian.

Carina [117]: a centrally placed ridge.

Carnivorous [133]: meat-eating.

Cartilage [118]: a white flexible connective tissue.

Cast [6]: a replica of the original shape of a shell or other part of an organism.

Cellulose [9]: a carbohydrate which is the main constructional material of plants.

Cenozoic [26]: an era, a division of geological time between 65 million years ago and the present day.

Cephalon [101]: the headshield of trilobites.

Cephalopoda [77]: a class of the Phylum Mollusca consisting of animals with chambered shells.

Chelicerae [107]: paired appendages in members of the Phylum Chelicerata used for locomotion, and where modified into pincers, for defence and food gathering.

Chelicerata [107]: a phylum of arthropods that includes horseshoe crabs, spiders and sea scorpions.

Chengjiang biota [44]: a Cambrian fossil Lagerstätte from China containing many arthropods and sponges.

Chitin [9]: a hard material made of protein found in the carapaces of arthropods.

Chlorite [9]: a green iron magnesium aluminosilicate mineral.

Chondrichthyans [118]: a major group of fish made up of sharks.

Chromosome [56]: a strand of DNA containing sequences of genes, paired with another strand found in the nucleus of cells.

Chronostratigraphy [40]: the area of dating of rocks that is concerned with absolute dates.

Cirripedes [106]: barnacles, members of a class of the Phylum Crustacea.

Cirripedia [104]: a class of the Phylum Crustacea.

CITES legislation [16]: international law drawn up by the Convention on International Trade in Endangered Species of Wild Fauna and Flora that prohibits movement of threatened plants and animals.

Clade [46]: a group of organisms derived from a single origin, monophyletic.

Class [18]: a division used in classification of plants and animals that contains one or more orders.

Clay [12]: fine-grained material usually composed of clay minerals.

Clay minerals [9]: widespread group of aluminium phyllosilicate minerals usually derived from the weathering of rocks.

Climax community [34]: the diverse and last assemblage of organisms found in an environmental setting or niche after a considerable length of time.

Cnidaria [68]: a phylum containing the corals, sea anemones, comb jellies, hydroids, and jellyfish all linked through the possession of cnidoblasts.

Cnidoblasts [68]: stinging cells in cnidarians used for protection.

Coccolithophore [57]: a unicellular planktonic organism covered by coccoliths.

Coccoliths [57]: minute circular plates composed of calcite that surround coccolithophores.

Coevolved [26]: referring to the evolution at the same time of two groups of organisms, i.e. insects and flowering plants.

Collagen [6]: a protein found in graptolite periderm and cartilage.

Colonial [72]: comprising groups of individuals living in close association and integration with others, as in bryozoans, graptolites and many corals.

Columnals [96]: the individual segments (ossicles) that make up the stem of a crinoid.

Commissure [89]: the line produced by the junction between two valves of a brachiopod.

Communities [34]: groups of diverse organisms living in the same place.

Compressions [9]: fossils, particularly plants and graptolites, that have become flattened and thus preserved.

Concentration [42]: a large volume of fossils preserved together, as in Concentration Lagerstätten.

Conchology [76]: the scientific study of shells.

Concretions [46]: nodular precipitations often made of siderite that in some cases may entomb organisms and fossilise them (as in the Mazon Creek Lagerstätte).

Condensation deposits [42]: assemblages of fossils found in caves.

Conodont apparatus [114]: the assemblage of elements in the anterior part of the conodont animal.

Conodonts [114]: phosphatic tooth-like structures found in conodont animals, members of a distinctive phylum that lived during the Palaeozoic. Used for biostratigraphical correlation and dating.

Conservation [42]: a term used for Lagerstätten where fossils display exceptional preservation, i.e. the Burgess Shale.

Consumers [34]: animals in a food-chain that feed on other organisms.

Continental shelf [33]: the area of shallow water to a depth of 200 m around the margins of continents.

Continental slope [34]: the steep area from the edge of the continental shelf to the abyssal floor of the ocean.

Coprofagous [98]: 'faeces-eating'. Some gastropods feed on the waste products of crinoids.

Coprolites [140]: fossilised faeces. Study of coprolites of fish, marine reptiles and dinosaurs can reveal useful information about diet.

Coquina [33]: a high concentration of shells that are preserved together.

Corallite [70]: the skeleton of a single polyp in a coral colony.

Core [22]: the central portion of the Earth at 2900 km deep, largely composed of iron and nickel.

Correlation [116]: matching rock units of the same time across geographical distances.

Cortical layer [113]: the outer wall of graptolite **rhabdosomes** (skeleton) made up of irregular bands of organic material that cover the **fusellar wall**.

Cosmopolitan species [74]: a plant or animal that is found in all parts of the globe rather than being restricted to certain environments or with a limited geographical range.

Cretaceous [26]: a division of geological time in the Mesozoic Era between 145 and 65 million years ago.

Crinoids [96]: members of a class of the Phylum Echinodermata, characterised by usually having a body on a stalk surrounded by arms that can be extended to form a filtration fan for capturing food.

Crust [22]: the outer layer of the Earth composed of basaltic rocks beneath the oceans and lighter rocks (igneous, metamorphic and sedimentary) that make up the continents.

Crustacea [9]: a diverse group of arthropods with a hard outer shell and jointed limbs; they include the lobsters, crabs, ostracods and barnacles.

Curation [15]: the process of preparing, labelling and numbering a geological specimen for storage in a museum.

Cyanobacteria [43]: a term for blue-green algae. They appeared early in Earth history and could photosynthesise, producing oxygen as a by-product.

Cynodonts [131]: a group of mammal-like reptiles that evolved during the Triassic with differentiated teeth and a hard palate.

Degassing [22]: the process in which gases are emitted into the atmosphere by volcanic activity.

Deionised water [13]: water that has been purified.

Dendroidea [110]: an order of graptolites comprising a bush-like skeleton with thecae of two types (cf. **Graptoloidea**).

Dentary bone [131]: a large bone in the lower jaw of lobe-finned fishes and tetrapods.

Denticles [114]: small pointed expansions developed along some conodont elements.

Deposit feeders [34]: organisms such as crustaceans, echinoids, gastropods and annelid worms that feed on the organic matter at or on the sediment surface.

Devonian [25]: a division of geological time in the Palaeozoic Era between 416 and 359 million years ago.

Diagenesis [5]: the changes that a sediment undergoes as it lithifies to become a sedimentary rock; these include dewatering, compaction, and cementation.

Diapsid [123]: a skull with two temporal openings behind the eye socket. Characteristic of the reptilian groups, the dinosaurs, pterosaurs, lizards, crocodiles, and the birds.

Diatomite [57]: a silica-rich sediment composed almost entirely of diatoms.

Diatoms [56]: small microscopic algae.

Dinobirds [129]: animals displaying similarities between the dinosaurs and the birds found in Cretaceous rocks in China.

Dinosaurs [127]: a diverse and important group of reptiles that dominated terrestrial environments during the Mesozoic.

Diploblastic [68]: having two layers of tissue, as in the corals and sea anemones.

Dissepiments [75/110]: cross-bars that connect branches in bryozoan colonies and stipes in dendroid graptolite colonies.

Diversity [34]: a term used to indicate the numbers of taxa present at any point in the fossil record or at any given geographical location.

Dorsal [89]: the upper part of a body or shell.

Doushantuo Formation [43]: a Precambrian Lagerstätte in China that has yielded microscopic plants and embryos of various animals.

Durophagous [118]: having the ability to crush prey, as do some fish and crustaceans.

Ecdysis [102]: process of moulting of hard outer exoskeleton in arthropods. This is necessary to allow the animal to grow.

Echinodermata [92]: a phylum of animals displaying five-fold symmetry and possessing a complex internal water circulatory system. It includes the echinoids, crinoids and the starfish.

Echinoid [93]: free-living members of the class in the phylum Echinodermata made up of a rounded test of interlocking plates. They include regular and irregular forms.

Ectoderm [68]: the outer layer of tissue in a diploblastic (twin-layered) organism.

Ectoplasm [63]: the outer layer of soft tissue in foraminiferans.

Ediacaran [23]: a division of geological time in the Upper Proterozoic between 635 and 542 million years ago.

Ediacaran biota [43]: exceptional and rare soft-bodied organisms such as jellyfish and other unusual forms dating from the Precambrian, found in Australia, Newfoundland and elsewhere.

Edrioasteroid [96]: a class of primitive members of the Phylum Echinodermata.

Eleutherozoans [92]: the free-living members of the Phylum Echinodermata that includes the echinoids, edrioasteroids, starfish and brittlestars.

Endoderm [68]: the inner layer of tissue in a diploblastic (twin-layered) organism.

Endolithic [86]: organisms such as sponges and bivalves that are capable of boring into stone.

Endoplasm [63]: the outer layer of soft tissue in the microscopic foraminiferans and radiolarians.

Endosymbiotic [64]: living within the body of another organism for mutual benefit.

Endothermy [128]: being warm-blooded.

Enteron [68]: the body cavity in the corals and sea anemones.

Eocene [41]: a subdivision (epoch) of the Paleogene period 55.8 to 33.9 million years ago.

Eon [22]: a major division of geological time such as the Phanerozoic that contains the smaller units, Eras.

Epibionts [81]: organisms that encrust the hard-parts of others, i.e. bryozoans on trilobites and cephalopods.

Epifaunal [34]: living on top of the sediment.

Epizoan [46]: organisms that live attached to other organisms.

Epochs [40]: a subdivision geological time within a geological period.

Era [25]: a division of geological time such as the Mesozoic that consists of a number of geological periods.

Eukaryotic [23]: cells with a distinct nucleus surrounded by a nuclear membrane.

Euryapsid [123]: a skull with a small opening high up behind the eye socket, such as that in the marine reptiles, the ichthyosaurs and plesiosaurs.

Eutheria [132]: a suborder of the Class Mammalia that is made up of the placental mammals.

Evolution [54]: the modification of organisms over time.

Exhalent siphons [84]: tubes in burrowing bivalves through which water is expelled (cf. **inhalent siphons**).

Exoskeleton [101]: having a skeleton on the outside of the body as in crustaceans.

Facial suture [101]: in trilobites the line of weakness that crosses the cephalon, which breaks open during ecdysis.

Family [18]: a division used in classification of plants and animals that contains one or more genera.

Fenestrules [72]: openings in meshwork bryozoan colonies delineated by the branches and **dissepiments**. They allow water to flow through the colony.

Filter feeders [89]: animals such as blastoids, brachiopods, bryozoans, and crinoids that feed by using structures such as lophophores or brachials, to filter out food from the sea water.

Flint [67]: a pale-coloured siliceous rock that produces a sharp edge when broken.

Foraminifera [62]: microscopic unicellular animals with ornate tests that live in a variety of marine and brackish water environments. They are important for biostratigraphical and palaeoclimate studies.

Form Genus [114]: a taxonomic level above species that has been applied to a disarticulated portion of an organism. Especially applied to different conodont elements and Pennsylvanian plants.

Form Species [114]: a taxonomic level below that of genus applied to disarticulated portions of organisms.

Form taxa [19]: taxonomic groups at any level to which form names have been applied.

Fossil assemblage [6]: the fossils now found preserved together in a particular unit.

Fossil Lagerstätte [42]: an example of well-preserved fossil material, either in great quantity, or exhibiting exceptional preservation.

Free-living [34]: organisms that are not attached to the substrate or to another organism.

Fusellar wall [113]: the inner wall of graptolite **rhabdosomes** (skeleton) made up of half-rings of organic material. Covered by the **cortical layer**.

Gametes [68]: sex cells such as sperms and eggs that, when fused, produce the offspring.

Gastropoda [82]: a class of marine and terrestrial coiled organisms within the phylum Mollusca.

Genal angle [102]: the angle produced by the lateral margin of a trilobite cephalon and its posterior margin.

Genus [18]: a taxonomic level used in the classification of plants and animals that contains one or more species.

Gill arches [117]: bony or cartilage supports of the gills in fish.

Gills [78]: Complex structures found in a variety of animals including trilobites, cephalopods and fish; used in respiration to extract oxygen in sea water.

Glabella [19]: the expanded bulge at the anterior of the cephalon in trilobites beneath which is situated the stomach.

Glaciation [35]: a cold period marked by the advance

of ice caps and ice sheets from polar regions and mountainous areas.

Globular [63]: spherical in shape.

Glossopteris flora [60]: plants that are part of a distinctive vegetation found in Gondwana.

Gnathostomes [117]: the first group of jawed fish that appeared in the Devonian.

Gonatoparian [101]: a style of facial suture in trilobites where it terminates at the genal angle.

Gondwana [132]: a large super-continent that existed in the Palaeozoic era, comprising South America, Africa, India, Australia and Antarctica.

Graptolites [110]: a class of colonial organisms which built skeletons of organic material and which were important zone fossils in the Palaeozoic.

Graptoloidea [110]: an order of graptolites comprising a simplified fret-like skeleton with thecae in which only one type of zooid, the autozooid, lived (cf. **Dendroidea**).

Gymnolaemata [74]: a class of bryozoans in which the individual animals live in box-like chambers in the colony.

Gymnosperms [60]: a diverse group of seed-bearing plants, including the conifers and the cycads, which were dominant in the Mesozoic.

Hadean [22]: an informal name for the division of time between the formation of the Earth 4567 million years ago and 4000 million years ago.

Hard palate [131]: a bony roof situated on the upper part of the mouth cavity.

Hemichordate [113]: animals that possess similar morphological features to chordates except that they lack a stiffening notochord.

Herbivorous [133]: plant-eating animals.

Hermaphrodite [68]: having both types of sexual organs in the one organism.

Hermatypic [71]: a group of corals that have a symbiotic relationship with photosynthetic algae which enhances growth rates. Known as the stony corals, they are important reef-formers.

Heteromorphs [81]: in ammonoids, unusually shaped shells that are radically different to normal coiled forms.

Holaspid [104]: the adult stage of trilobites reached after all phases of moulting have been completed.

Holdfast [67]: a root-like structure at the base of crinoid stems that firmly holds the animal in place in the sediment.

Holotype [18]: the primary specimen used for the recognition of a new species.

Homalozoans [100]: unusual and rare group of organisms with similarities to the echinoderms, found during the Palaeozoic, and consisting of a head

with one 'arm' and a tail.

Hominids [134]: members of the Family Hominidae, which includes the great apes – gorillas, chimpanzee, orang-utans and modern humans.

Hominins [134]: members of the Tribe Hominini, which includes the chimpanzees and modern humans.

Horizon [12]: a level within a sequence of rock, which may be a thin unit within a bed, a bed itself, or a collection of beds.

Hydrospires [98]: delicate folds of tissue in blastoids that were used for respiration.

Hydrothermal [46]: related to the heating up of water within the Earth's crust. One such location is at mid-oceanic ridges where hydrothermal vents are colonised by unusual animals.

Hyponome [78]: a tube situated beneath the mouth of cephalopods through which water is expelled at speed to facilitate swimming.

Ichnofacies [140]: a term used for a unit characterised by a particular suite of trace fossils that indicate the nature of the geological setting and biological activity of the organisms that lived there.

Ichnogenus [19]: a taxonomic rank above that of ichnospecies, applied to trace fossils.

Ichnology [137]: the science of the study of trace fossils.

Ichnospecies [19]: a taxonomic rank below that of ichnogenus, applied to trace fossils.

Ichthyologist [119]: a scientist who studies fish.

Index fossil [36]: a fossil that has a short range and a wide geographical distribution and which is diagnostic of a particular geological unit. Graptolites, cephalopods, and some trilobites have been employed as index fossils in biostratigraphy.

Infauna [76]: animals that live within the sediment.

Inhalant [72]: inward moving, such as water currents generated by the beating of cilia in lophophores in bryozoans.

Inhalent siphons [84]: tubes in burrowing bivalves and gastropods through which water is taken in (cf. **exhalent siphons**).

Interambulacra [93]: the five sections of the test (shell) in echinoids between ambulacra.

Invertebrate [76]: lacking a backbone.

Iron pyrites [14]: an iron sulphide mineral FeS_2 that is found in many geological settings including in anoxic conditions in muds and deep-water sediments, and which develops cubic bronze-coloured crystals; commonly called 'Fool's Gold'.

Jurassic [40]: a division of geological time in the Mesozoic Era between 200 and 145 million years ago.

Juveniles [34]: young animals that have not reached

sexual maturity.

Karstic [42]: pertaining to erosion features produced by flowing water, such as clints, grykes, and caves developed in limestone.

Kingdom [18]: the highest division of life used in classification.

Lagerstätte [42]: a German term for a deposit containing material of economic value.

Lappets [81]: lateral projections of the shell in some ammonites which may have protected the tentacles.

Lazarus Taxon [60]: an organism that appears to disappear from the geological record, only to reappear. The gap may be explained due to it simply not having been preserved, or its non-discovery by palaeontologists, in that intervening time.

Lectotrophic [74]: a type of larva that settles close to their parents.

Leucon [65]: the most complex arrangement of sponge architecture, with water canals interspersed with collar cells.

Ligaments [84]: tough tissue used to connect muscle to bones, or shells together, such as in bivalves.

Limestone [12]: a sedimentary rock composed of calcium carbonate, deposited usually under marine conditions, and composed of chemically-derived lime muds or fossilised shells, or both.

Lithified [6]: having been cemented together; sediments turned into solid rock.

Lithologies [114]: referring to the physical characteristics of different rocks.

Lithostratigraphy [40]: the definition of units of rocks in terms of their lithology and their correlation from area to area.

Living fossil [26]: a term applied to organisms that have a very long range and which are still found living today. Some examples include the brachiopod *Lingula*, the cephalopod *Nautilus* and the coelacanth fish *Latimeria* and the plant *Ginkgo*.

Lobes [79]: the pointed portion of the suture in ammonoids that point away from the body chamber.

Longirostrine [128]: having a long snout, as in some crocodilian groups.

Lophophore [72/89]: in bryozoans the tentacle crown used to generate incoming water currents for feeding; in brachiopods a spiral arrangement of tissue used for feeding.

Lumen [96]: a hole situated in the centre of ossicles in crinoids, through which is found a thread of flesh.

Ma [22]: abbreviation meaning 'millions of years'.

Macrofossil [13]: a fossil that is large enough so that it can be seen with the naked eye.

Madreporite [93]: a plate containing small holes on the dorsal surface of echinoids through which water

is taken into the water-vascular system.

Malacology [76]: the study of shells.

Malacostraca [104]: a class in the phylum Crustacea containing shrimps, lobsters and crabs.

Mammal-like reptiles [131]: reptiles that possessed features such as varied dentition, and which resembled later more advanced mammals.

Mantle [22/76]: (1) the central layer in the Earth between the crust and core (between 50 and 2900 km) composed of dense ultrabasic rocks; (2) the tissue in molluscs and brachiopods that is responsible for secreting the shell.

Mantle cavity [76]: a fleshy cavity in molluscs containing some of the vital organs such as the gills.

Manus [121]: the fore limb in tetrapod animals.

Marine [33]: relating to the oceans.

Marl [47]: a clay or muddy material composed of calcareous material.

Marsupials [48]: a group of mammals that give birth to their young prematurely and brood them in a pouch until fully mature.

Mass Spectrometer [15]: a precision instrument that can measure the chemical composition of materials.

Mass-mortality [120]: many individuals or colonial organisms dying at the same time.

Maturity [60]: the degree to which a sedimentary rock has been heated.

Meraspid [104]: the intermediate stage during the growth sequence in trilobites during which most of the thoracic segments are inserted.

Mesenteries [68]: infolded layers of tissue in the polyps of corals. These increase the surface area for food digestion and gas exchange.

Mesodermal [92]: a layer that is sandwiched between two other layers, such as the test or shell in members of the Phylum Echinodermata.

Mesozoic [26]: an era, a division of geological time between 251 and 65 million years ago.

Metatheria [132]: a suborder of the Class Mammalia that is made up of the marsupials.

Microfossil [14]: a fossil that is so small that it can only be seen with the aid of a microscope.

Mineralisation [23]: the process of producing a hard shell or skeleton by the precipitation of minerals.

Miocene [41]: a subdivision (epoch) of the Neogene period 23 to 5.3 million years ago.

Mississippian [26]: a division of geological time in the Palaeozoic Era between 359 and 318 million years ago; equivalent to the now informal term 'Lower Carboniferous'.

Mollusca [76]: a diverse phylum containing cephalopods, gastropods and bivalves lined by the possession of a muscular 'foot'.

Monotremes [132]: a primitive group of egg-laying mammals that include the Duck-billed Platypus and the Echidnas.

Monticles [72]: elevated portions of some bryozoan colonies that enabled the ejection of waste materials via exhalent currents.

Monticulate [75]: the state of having monticles.

Morphospecies [19]: a fossil species defined on the basis of the morphology, rather than on genetic information and biological behaviour, as are modern species.

Mosasaurs [78]: marine reptiles with large skulls and teeth that were major predators in the Late Cretaceous.

Mould [6]: the cavity remaining in sediment or rock after a shell or skeleton dissolves away.

Mud [12]: a very fine-grained sediment that lithifies into mudstone or shale.

Multicellular [23]: composed of many cells.

Neogene [26]: a division of geological time in the Cenozoic Era between 23 and 2.6 million years ago.

Niche [34]: the role or situation of an organism in its environment.

Nocturnal [134]: organisms that are active at night, i.e. early mammals, bats, and prosimians.

Notochord [115]: an elongate rod running alongside the spinal cord in primitive chordates which supports the body. It developed into the backbone in later vertebrates.

Nucleus [23]: the central portion of a cell containing the DNA.

Obrution deposit [43]: an accumulation of fossils that is well preserved due to its being buried rapidly under sediment, perhaps moved following a storm event.

Oil shale [48]: a fine-grained muddy sedimentary rock containing a high proportion of hydrocarbons.

Opaline [8]: resembling opal, a milky translucent mineral, in appearance.

Operculum [82]: a skeletal lid that closes off the body chamber in gastropods or the zooecial chamber in some bryozoans.

Ophiuroidea [99]: the brittlestars, a group of free-living members of the Phylum Echinodermata.

Opisthoparian [101]: a style of facial suture in trilobites where it terminates at the posterior margin of the cephalon.

Opisthosoma [107]: the abdomen in the Chelicerates such as the eurypterids, horseshoe crabs and the spiders.

Order [18]: a division used in classification of plants and animals that contains one or more families.

Ordovician [25]: a division of geological time in the Palaeozoic Era between 488 and 444 million years ago.

Ornithischian [125]: one of two major divisions of the dinosaurs whose two pelvic bones, the pubis and the ischium, both point backwards; 'bird-hipped'.

Orthoconic nautiloids [46]: cephalopods with straight shells and simple sutures.

Ossicles [92]: the individual segments (columnals) that make up the stem of a crinoid.

Ostia [65]: tiny opening in the wall of sponges through which water is taken into the body cavity.

Ostracod [106]: a small crustacean that lives in a bean-shaped pair of shells and is found in many environments. Useful biostratigraphical marker.

Ostracoda [104]: a class in the Phylum Crustacea.

Ostracoderms [117]: the oldest fish; these were jawless and appeared during the Cambrian.

Palaeoecology [10]: the study of the relationship of fossilised plants and animals and of plant or animal to each other and to the environment in which they once lived.

Palaeogeography [21]: the distribution of continents and oceans over geological history.

Palaeozoic [25]: an era, a division of geological time between 542 and 251 million years ago.

Paleogene [26]: a division of geological time in the Cenozoic Era between 65 and 23 million years ago.

Pallial line [84]: a line near the anterior margin on the interior of bivalves that marks the outer limit of the attachment of the mantle.

Palynologists [37]: scientists who study palynomorphs.

Palynomorph [37]: fossil spores, seeds, and pollen.

Pangaea [123]: a super-continent composed of Gondwana to the south and Laurasia (North America, Europe and Asia) to the north that existed from the Permian until its breakup during the Cretaceous.

Parallel-evolution [133]: development of similar morphological features in two different biological groups, sometimes at the same time.

Paratypes [18]: specimens other than the holotype that form the basis of a new species.

Part and counterpart [12]: two halves of a fossil when split open.

Patelliform [83]: having a shell shaped like a hat, as in the modern limpet.

Pedicle [89]: a fleshy stalk used to attach some groups of brachiopods to the substrate.

Pedicle foramen [89]: a small hole in the valves of some brachiopods through which extends a fleshy foot, the pedicle.

Pelagic [34]: a term used for animals that live in the upper waters of the ocean.

Pelmatozoans [92]: attached members of the Phylum

Echinodermata such as the blastoids, crinoids and cystoids.

Pennsylvanian [40]: a division of geological time in the Palaeozoic Era between 318 and 299 million years ago; equivalent to the now informal term 'Upper Carboniferous'.

Pentameral [89]: having five of something.

Pentameral symmetry [96]: five-fold symmetry, as exhibited in the members of the Phylum Echinodermata.

Periderm [6]: the outer layer of graptolite rhabdosomes.

Period [21]: a division of geological time such as Cambrian, Silurian and Triassic that contains epochs.

Periproct [93]: the anus in echinoids.

Peristome [93]: the mouth in echinoids.

Permian [25]: a division of geological time in the Palaeozoic Era between 299 and 251 million years ago.

Permineralisation [6]: the process by which small pores in shells, bones or plant tissue become infilled with minerals that precipitate from percolating water.

Pes [121]: the hind limb in tetrapod animals.

Petrological microscope [15]: a microscope adapted for studying thin-sections of rocks, minerals and fossils.

pH [42]: a measure of the acidity or alkalinity of a solution.

Phanerozoic [23]: an Eon, a major division of time between 542 million years ago and the present day.

Phloem [6]: the tubular vascular tissue in plants responsible for the transportation of sugars from the leaves.

Photic zone [34]: the upper part of a water body in which light can penetrate; this is usually to a depth of 100 m.

Photosynthesis [34]: the process whereby plants can produce sugars from water and carbon dioxide under sunlight. An important byproduct of this process is oxygen.

Phragmocone [81]: the shell of cephalopods.

Phylactolaemata [74]: a class of freshwater bryozoans that have no hard parts and consequently a poor fossil record.

Phylogeny [21]: the evolutionary history of a taxonomic group.

Phylum [18]: a taxonomic rank beneath that of Kingdom and above that of Class.

Pinnules [97]: small feathery expansions found on the brachials of crinoids that increase the efficiency of the filtration fan for collecting food.

Pioneer community [34]: an assemblage of animals and plants that are the first to inhabit an area recently made available. Deposition of sediment can produce a barren bedding plane, which is then exploited.

Placental mammals [132]: a group of vertebrates that give birth to their offspring live. In the womb, the foetus is connected to its mother by means of a placenta through which nutrients and waste materials are passed.

Placoderm [118]: primitive heavily armoured fish with jaws that lived between the Silurian and Devonian periods.

Planctotrophic [74]: a type of larvae with cilia and a food source that can travel great distances, and so far away from their parents.

Planispiral [63]: a flattened coiled shell as in some gastropods, cephalopods and foraminiferans.

Plankton [57]: a range of micro-organisms that live in the uppermost waters of the oceans.

Planktonic [64]: living in the surface waters of the oceans.

Pleistocene [41]: a subdivision (epoch) of the Quaternary period between 2.5 million years and 12,000 years ago.

Plicated [89]: a term for folded or corrugated, as in the commissure of rhynchonellid brachiopods.

Pliocene [41]: a subdivision (epoch) of the Neogene period 5.3 to 2.5 million years ago.

Polyphyletic [101]: relating to different phyla.

Polypide [72]: the individual animal in bryozoans and some other colonial groups.

Polyps [68]: the attached soft-tissued part of the corals and sea anemones.

Porifera [65]: the phylum comprising the sponges.

Porous [63]: having small holes through which gas and liquids can travel.

Posterior, -ly [77]: the back end of a body or shell.

Predator [34]: an animal that preys on others.

Primates [134]: an advanced order of mammals with hands, nails, and forward-facing eyes that enable stereoscopic vision.

Producers [34]: organisms in the food-chain that can produce energy through photosynthesis.

Progrades [33]: grows into, as in a delta developing at the mouth of a river into the ocean.

Prokaryotic [23]: a primitive cell that lacks a defined nucleus or additional organelles.

Proparian [101]: a style of facial suture in trilobites where it terminates at the lateral margin of the cephalon.

Prosimians [134]: a primitive group of primates that include the lemurs.

Prosoma [107]: the fused head and thorax in the Chelicerates such as the eurypterids, horseshoe crabs and the spiders.

Protaspid [104]: the first larval stage in the ontogeny (growth sequence) of the trilobites.

Proterozoic [23]: an Eon, a major division of time between 2500 and 542 million years ago.

Protista [56]: a Kingdom containing both animal and plant single-celled organisms with a cell nucleus. Includes the foraminiferans, radiolarians and the diatoms.

Protoconch [77]: the first shell secreted by the larva; sometimes preserved in cephalopods and some other molluscs.

Proto-continents [22]: small landmasses that developed early in the Earth's history. Convergence of many such masses led to the generation of the larger continents.

Provincialism [34]: distribution of plants and animals into discrete geographic provinces. During the Lower Palaeozoic, trilobite assemblages exhibited provincialism, and they have aided the understanding of the palaeogeography of that time.

Pseudopodia [63]: the ectoplasm in foraminifera that can be extruded beyond the test and used to capture prey.

Pterobranch [113]: a small colonial worm-like animal that secreted tubes in which to live. Considered to be hemichordates and thought to be similar in morphology to the graptolites.

Pygidium [101]: the tail of trilobites, which may display pseudosegmentation or may be smooth.

Pyriform [63]: a term used for 'pear-shaped'.

Pyrite disease [14]: a term applied to pyrite that has undergone severe oxidisation. As it does so, the mineral disintegrates as well as producing sulphuric acid.

Quadrupedal [126]: organisms that stand on all four limbs.

Radial symmetry [68]: symmetry around a central point as in scleractinian corals.

Radiation [46]: the evolution of plants or animals from a single starting point into new and diverse taxa.

Radiolarians [62]: microscopic animals with tests composed of silica that form part of the plankton in the uppermost parts of the oceans.

Radula [83]: a hard tongue-like structure in gastropods used by them to browse hard surfaces for nutrients.

Rancho La Brea [50]: an area in Los Angeles, California where abundant examples of Pleistocene animals were preserved in asphalt tar-pits.

Recrystallised [9]: a material that has been altered by pressure so that it dissolves before forming new crystals.

Reef-builders [71]: organisms such as many corals, whose skeletons contribute to reefs that build up as discrete mounds or ridges from the seabed.

Regionally Important Geological and Geomorphological Sites [16]: These are sites of scientific interest, but which are considered to be less important than Sites of Special Scientific Interest (SSSI) and so have no statutory protection. Many are maintained and documented by local volunteer groups.

Regressions [30]: periods when the sea level of the oceans fall and continental shelf areas may become exposed.

Reticulate [72]: having a meshwork colony form, as have many bryozoans.

Rhabdosome [110]: the skeleton of graptolites composed of organic materials.

Rheophobic [98]: a term used for organisms that avoid water currents, such as those found at abyssal depths.

Rheophyllic [98]: a term applied to organisms that can tolerate water currents.

Rhynie Chert [46]: an unusual siliceous rock containing exceptionally preserved early Devonian plants and some animals from northeast Scotland.

Rudists [84]: odd-shaped bivalves common in the Mesozoic that were important reef-builders.

Rugosa [69]: a suborder of Palaeozoic corals, both solitary and colonial, with conspicuous septae.

Saddle [79]: the rounded portion of the suture in ammonoids that points towards the body chamber.

Sarcodina [62]: a phylum of amoeba-like microorganisms that includes the foraminiferans.

Sarcopterygiians [118]: a major group within the bony fishes characterised by having paddle-shaped fins which were the ancestors of the first terrestrial animals. Informally called the lobe-finned fishes.

Saurischian [125]: one of two major divisions of the dinosaurs whose two pelvic bones, the pubis and the ischium, point in opposite directions; 'lizard-hipped'.

Sauropods [125]: members of the suborder Sauropoda – a group of large herbivorous dinosaurs, such as *Diplodocus*, which were quadrupedal.

Scales [117]: small bony or horny plates found covering the skin of fishes and reptiles. These provide protection.

Scanning Electron Microscopes [25]: highpowered instruments that use beams of electrons to produce very high magnification views of objects. These views can be captured using cameras attached to these microscopes.

Scleractinia [71]: a suborder of post-Palaeozoic corals that are either solitary or colonial. The hermatypic corals are associated with symbiotic algae and are important reef-building forms.

Sclerotised [101]: having chitin in the exoskeleton of some arthropods hardened by the addition of calcium carbonate or calcium phosphate.

Sedimentological [33]: the processes and characteristics relating to sediments and the rocks that they form.

Septae [69]: In cephalopods a skeletal partition that separates chambers from each other; in corals they are vertical plates that add strength to the corallite.

Series [40]: the rocks contained within an Epoch.

Sessile [34]: meaning attached to the substrate. Sessile organisms include bryozoans and stalked crinoids amongst others.

Sexual dimorphism [81]: having different morphology depending on sex. In many ammonite species the females are larger than the males.

Siderite [47]: an iron carbonate mineral often found in concretions.

Silica [5]: an oxide of silicon that is found in a variety of forms, and which is utilised by some organisms such as sponges, radiolarians, and diatoms to form skeletal elements.

Siliceous [9]: being composed of silica.

Silurian [41]: a division of geological time in the Palaeozoic Era between 444 and 416 million years ago.

Siphuncle [77]: a thin strand of soft tissue that connects all the chambers in the cephalopods.

Sites of Special Scientific Interest [16]: a title given to areas that have yielded scientifically important materials and information, which have been given legal protection from damage and collecting.

Small Shelly Fossils [25]: a collective term for the remains of the earliest shelly organisms that appeared at the beginning of the Cambrian.

Species [18]: a taxonomic rank below that of genus, comprising a naturally occurring interbreeding population.

Spicules [67]: small spines of various shapes made of silica that are found in the walls of sponges and provide some support to the soft tissue.

Spores [58]: a unicellular reproductive cell produced by lower plants such as fungi and mosses.

Stable isotopes [35]: an isotope that is not radioactive. Isotopes are atoms of a chemical element that differ from each other in the number of neutrons contained in their nucleus.

Stage [40]: a small unit of rock contained within a short subdivision of geological time.

Stenolaemata [74]: a class of bryozoans in which the zooids live in tubular chambers.

Stereoscopic vision [134]: vision that allows for the perception of depth through the superimposition of two slightly different views from each eye.

Stipes [110]: branches in dendroid graptolites.

Stolon [110]: a thread of flesh that runs through graptolite colonies connecting individual animals.

Stomata [6]: openings on the underside of leaves of angiosperms that allow for the diffusion of gases.

Stratigraphy [39]: the discipline in geology concerned with the ordering of the rock succession and its dating.

Stromatolites [23]: small buildups, comprising layers of cyanobacteria and trapped sediment that developed in shallow marine waters. Early examples are found in Precambrian rocks, and modern forms in Western Australia.

Subspecies [18]: a subdivision or lower taxonomic rank of a species.

Substrate [71]: the surface (often the seabed, rocks, or shells) on which an organism will settle and grow.

Succession [10]: the sequence of rocks within a particular geological time unit.

Surfactant agent [13]: a chemical that reduces the surface tension of liquids. It helps to disperse oily materials and is used for cleaning.

Suspension feeders [34]: animals that feed on microscopic particles of food floating (suspended) in the surrounding water.

Suture [78]: in cephalopods, the line where the septa meet the outer wall.

Symbiotic [71]: having a relationship in which two organisms live in close proximity to each other and mutually benefit from the arrangement.

Sycon [65]: a body form in sponges where it is slightly invaginated.

Synapsid [123]: a skull with a single large temporal opening behind the eye socket, as in mammal-like reptiles.

Synecology [34]: the study of the relationship of communities of plants and animals to each other and to the environment in which they live.

Systems [40]: a major division of the geological rock succession equivalent to the geological period used for the division of that time, i.e. Cambrian System – the rocks of the Cambrian Period.

Tabulae [71]: horizontal partitions in the calyx of corals. Particularly well-developed in the Palaeozoic suborder Tabulata.

Tabulata [71]: a suborder of colonial Palaeozoic corals that have poorly developed septa and frequent tabulae.

Taphonomy [5]: the subdivision of palaeontology concerned with the processes of fossilisation.

Tar-pits [50]: depressions containing accumulations of hydrocarbons such as asphalt; those in Los

Angeles were the site of exceptional preservation of a Pleistocene biota.

Taxon [10]: a taxonomic rank at any level, i.e. class, order, genus or species.

Tectonic activity [9]: the movement of lithospheric plates.

Tegmen [97]: a covering made up of several plates over the top of the theca (or cup) in crinoids.

Tertiary [26]: a now obsolete name of a geological period in the Cenozoic era, replaced with Paleogene and Neogene.

Test [92]: the skeleton of echinoids, and of foraminiferans.

Tetrapod [121]: an animal with four legs; a higher taxonomic group that includes amphibians, reptiles, birds and mammals.

Theraspids [131]: a group of Permian reptiles that developed some mammalian features.

Theropods [125]: members of the suborder Theropoda, a major group of bipedal saurischian dinosaurs.

Thin-section [15]: a slice of a rock that is glued between glass, and cut and ground down thin enough to allow light to pass through. They would be examined using a petrological microscope. Microfossils in limestone can be viewed by this technique.

Thoracic breathing [123]: using the muscles of the ribcage to aid the expansion of the lungs and facilitate breathing.

Thorax [101]: the segmented middle portion of the body in many arthropods.

Tooth-and-socket [84]: a series of interlocking nodes and cavities along the hinged margin of some bivalves, which strengthens the point of articulation.

Tracks [121]: imprints made in sediment by an animal with legs.

Trackway [121]: a pathway made by an animal.

Trail [138]: the drag-marks made in sediment by an animal without legs.

Triassic [26]: a division of geological time in the Mesozoic Era between 251 and 200 million years ago.

Trilobita [101]: a major phylum of marine arthropods that dominated Lower Palaeozoic ecosystems. They became extinct at the end of the Permian.

Trochiform [83]: a pyramidal-shaped shell in gastropods.

Trophic level [34]: the position within a food-chain or food-web of an ecosystem.

Tube feet [93]: elongate water-filled expansions of soft tissue found in echinoderms. In echinoids they protrude from the pores on ambulacra and in starfish from the underside of arms, and are used to aid movement, respiration and feeding.

Turbidity currents [45]: water currents that move down the continental slope carrying sediment.

Ultrasonic tank [13]: a water tank that vibrates using ultrasonic waves. It is often used to clean fossils as the waves shake off tiny particles of adhering sediment.

Umbo [84]: the pointed beak of a bivalve shell situated on the dorsal margin.

Umbonal angle [86]: the angle prescribed by the posterior and anterior sides of a bivalve shell measured either side of the umbo.

Unicellular [56]: composed of one cell.

Uniramia [108]: a phylum of arthropods containing the insects as well as centipedes and millipedes.

Uniramous [108]: a limb that is not divided into two morphological parts, unlike that of trilobites.

Valves [84]: the individual shells of an organism.

Vascular plants [57]: advanced plants with tubes (xylem and phloem) in their stems used to carry nutrients, sugars and water.

Ventral [89]: the lower part of a body or shell.

Visceral mass [84]: the soft tissue in molluscs containing the vital organs.

Xylem [6]: the tubular vascular tissue in plants responsible for the transportation of water and dissolved nutrients from the roots.

Zone fossils [36]: fossils that define a narrow time period known as a zone.

Zone [40]: a short-ranged unit of geological time containing rocks with distinctive zone fossil.

Zooids [72]: individual animals in bryozoans; some are primarily feeders while others may be modified for defence, brooding or cleaning duties in the colony.

Zooxanthellae [71]: algae that can photosynthesise. Some are symbiotic with corals and radiolarians.